1991

Daytime Television Programming

ELECTRONIC MEDIA GUIDES

Daytime Television Programming

Marilyn J. Matelski

Boston College

Ⓕ

Focal Press
Boston   London

Focal Press is an imprint of Butterworth–Heinemann.

Recognizing the importance of preserving what has been written, it is the policy of Butterworth–
Heinemann to have the books it publishes printed on acid-free paper, and we exert our best efforts to that end.

Library of Congress Cataloging-in-Publication Data
Matelski, Marilyn J., 1950–
     Daytime television programming / Marilyn J. Matelski
          p.   cm. —(Electronic media guide)
     Includes bibliographical references.
     ISBN 0-240-80087-7
     1. Television programs—United States.   2. Television
     broadcasting—United States—Management.
     I. Title. II. Series.
     PN1992.3.U5M27   1991
     791.45'0236'0973—dc20
                                        90-45630
                                      CIP

British Library Cataloguing in Publication Data
Matelski, Marilyn J. *1950–*
     Daytime television programming.—(Electronic media
     guides)
     1. United States. Television programmes. Scheduling
     I. Title II. Series
     791.450973

     ISBN 0-240-80087-7

Butterworth–Heinemann
80 Montvale Avenue
Stoneham, MA 02180

10 9 8 7 6 5 4 3 2 1

Printed in the United States of America

For Mom—my proofreader, my advisor, my friend

Contents

Introduction

Television (TV) executives have always valued the daytime programming slot as a major ingredient in broadcast scheduling, both nationally and locally. This is because for the most part, daytime television is still the best deal in the business—serving large audiences at comparatively low costs. In fact, it can be said that daytime television actually helps to support prime-time TV because of its impressive record of investment versus yield.

In reality, most local programmers (whether independent or network-affiliated) are heavily involved in daytime show decision-making—perhaps more than any other period of their broadcast day. For this reason, it is important to consider all aspects of this daypart.

Accordingly, Chapter 1 will present a brief history of daytime television in the United States. Chapters 2 through 6 will address specific program genres—talk shows, game shows, soap operas, children's programming, and specialized fare—and compare them to their nighttime counterparts in terms of cost, audience, promotional appeal, etc. In the last chapter, specific daytime shows and their demographic values will be explored to illustrate the enormous challenge and creativity needed to run a successful local television station—especially during this very important time period.

1

▼ A Brief History
▼
▼
▼
▼

For many years, the major objective in American television programming was to provide a newer, faster, and more cost-efficient technology to serve the nation's viewing public. This preoccupation was due, in part, to an intense international competition (begun in the latter half of the nineteenth century) to be first in offering continuous broadcast services to interested audiences. In fact, by the late 1800s, many nations, including Russia, Germany, France, Great Britain, Japan, and the United States, had already begun the race against time to make their mark in media history. Thus, most nations during the early stages of television development viewed the video product, or *software*, as secondary to the primary goal of providing a stable signal as often and to as many viewers as possible (see Table 1).

As time went on, however, several of these nations began to concentrate more on programming content than on technical quality. They were now most concerned with the *use* of this newfound medium, not necessarily with developing other technological innovations (which they felt could be incorporated at a later date, anyway). To them, the critical issue was the message, not stylistic presentation. For example, countries such as Russia and Nazi Germany saw television from an *authoritarian* perspective; they felt that all forms of mass communication should be used as the information-education-propaganda distribution arm of the government. Other nations, like France and Great Britain, took a more moderate approach, giving governmental support (as well as limited governmental direction) to the broadcast airwaves. Their broadcast philosophy was referred to as a *benevolent* system—one which used the media to provide audiences with what they needed—namely, education, information and cultural taste—not necessarily with what they thought they wanted.

In the United States, however, the philosophical perspective behind programming was quite different. As mentioned earlier, many American broadcasters saw technological superiority as most necessary and appropriate in the overall development of television. In addition, they felt that the airwaves, while public, should not be subjected to any governmental influence. This point of view, the *democratic* perspective, was thus defined as the process of relying on a marketplace of free enterprise to determine the most useful programming to be shown on the airwaves. Based on this philosophy, television genres would be freely introduced to enthusiastic viewing audiences. Some shows would succeed, while others would fail. According to the U.S. programming point of view, the best television fare would survive— and improve with time.

Television was introduced officially to the American public by President Franklin Delano Roosevelt, as he announced (via the airwaves) the opening of the

▶ **Table 1** The international race for television: a capsule history, 1884–1964

1884	Paul Nipkow files for a patent in Berlin. His invention, the "electronic tele-scope," is a mechanical visual scanning device.
1897	Germany's Karl Braun invents the cathode ray tube.
1907	Russian scientist Boris Rozing files a patent, using the cathode ray tube as an electronic scanner.
1908	Great Britain's A.A. Campbell Swinton creates a visual image on a fluorescent screen via the cathode ray tube.
1924	Britisher John Logie Baird receives a patent for his development of a mechanical scanning system.
1925	American Charles Jenkins develops a mechanical scanning system much like the one developed by Baird.
1927	Philo T. Farnsworth develops a working electronic camera.
	Japanese inventor Kenjiro Tahayanagi demonstrates a laboratory transmission of visual signals.
	Vladimir Zworykin develops an electronic iconoscope camera for RCA.
1928	Hungary's Von Mihaly demonstrates a mechanical scanning system.
1931	Russian Semyon Katayev builds an electronic camera similar to Zworykin's iconoscope.
1935	Continuous television services begin in Germany.
1936	The British Broadcasting Corporation begins continuous television service to its customers.
1938	Moscow and Leningrad begin transmitting television signals.
1939	President Franklin Delano Roosevelt introduces television to the U.S. at the New York World's Fair.
1941	The Federal Communications Commission grants NBC and CBS commercial licenses for all their television stations.
1946	The British Broadcasting Corporation resumes its normal television services after being suspended during World War II.
	France nationalizes its television system, *Radiodiffusion Television Francaise* (RTF).
1947	RCA and Paramount Pictures experiment with television screenings in movie theaters.
late 1940s	Japan is granted permission from American post-war occupying forces to refor-mulate its television system, the *Nippon Hosai Kyokai* (NHK).
1950	The BBC successfully attempts its first cross-channel broadcast.
1952	Nippon Television Network (NTV), a commercial system, is granted permission to compete with Japan's NHK.
1953	Japan, Switzerland, and Belgium begin continuous national television services.
1954	The British parliament approves a commercial television system to compete with the BBC. It is called the Independent Television Authority (ITA).
1956	The first TV station originates programming in the Middle East.
	Italy operates its television system as a private monopoly, largely owned by the government. It is named *RAI-Radiotelevisione Italiana*.
1964	Political controversy in France causes the government to create a more autono-mous communications authority, the *Office de Radiodiffusion-Television* (ORTF).

1939 New York World's Fair. The spectators in attendance that day were instantly thrilled and delighted by the new medium; and despite the subsequent consumer distribution difficulties brought on by World War II, it was clear from the very beginning that TV viewership would soon surpass the large audiences previously attracted to radio. Castleman and Podrazik, in *Watching TV: Four Decades of American Television*, make this point most dramatically. According to their statistical findings of television set production from 1946-1947, the following profile emerged:

- From January to August, 1946, only 225 TV sets were produced in the U.S.
- During the month of September 1946, 3242 receivers were manufactured.
- By January 1947, the monthly TV set production figure had jumped to 5437.
- By May 1947, at least 8690 television sets were manufactured each month.[1]

In 1946, only 8000 television sets could be found in American homes. However, by 1950, over 4 million household receivers existed across the country. As noted by Linda J. Busby, author of *Mass Communication in a New Age*, the effect of increased TV set production was equally dramatic and immediate for TV broadcasters and advertisers, as well as for audiences:

> The public was eager to have local television stations established in their communities, and advertisers were eager to use the medium to market their products. Television was a success with advertisers because it almost immediately established an excellent track record. The lipstick manufacturer Hazel Bishop, for example, was doing an annual business of $50,000 in the early 1950s. During 1952 the company began advertising on television, and that year sales zoomed to the $4.5-million mark.[2]

As the technology of television continued to develop, so did the broadcasters' commitment to programming. In 1944, for example, the four networks in existence at the time—CBS, NBC, ABC, and DuMont—provided a combined total of only 16.5 prime-time entertainment hours for TV viewers. Just 2 years later, however, the number of prime-time programming hours increased by almost 55% to 25.5 hours. By 1952, each network was broadcasting at least 18 evening hours per week.[3] ABC, NBC, CBS, and DuMont also aired programming during daytime hours and continued to build station affiliateships throughout the country.

In the early 1950s, affiliated stations were extremely important to network survival. In addition to the obvious lure of higher advertising revenues through national spots, local stations often provided the talent and creativity needed for future network programming. For example, a Chicago celebrity named Dave Garroway began his television career on a 30-minute local show entitled "Remember the Days." He later hosted NBC's "Today" (among other ventures). Fran Allison and Burr Tillstrom originated a children's program, "Junior Jamboree," which later became "Kukla, Fran & Ollie" on NBC. Most notably, in 1950, a zany disc jockey, Ernie Kovacs, provided both talent and a creative show concept. He also demonstrated that network television could be successful in an early morning time slot:

> At the start of 1952, daytime TV programming was still exceedingly sparse. A few stations signed on at about 10:00 A.M., but nothing of any importance took place until about 4:00 P.M. One exception was WPTZ in Philadelphia which, each

weekday morning from 7:00 A.M. to 9:00 A.M., ran *"Three To Get Ready"* a loose show led by former radio disk jockey Ernie Kovacs. The program had begun in late November, 1950, and featured some live music, records, time and weather checks, and great doses of Kovacs' own peculiar television insanity. He read fan letters on the air, performed skits he had written himself, shot off toy guns after puns, picked his teeth, and even held an audition for goats. Oddly enough, *"Three To Gey Ready"* did very well in the local ratings and the success of Kovacs apparently convinced NBC that [producer Pat] Weaver's idea for an early morning show might attract network viewers as well.[4]

In point of fact, network executives soon began to realize that while prime-time television programs received the most attractive advertising revenues in the television business, they also were quite costly. Daytime shows, on the other hand, served comparatively large audiences at very low prices. Their impressive investment-versus-yield records often saved the networks from ledgers with red ink. In short, the profits from daytime television, along with the inherent flexibility of the timeslot to showcase new talent as well as innovative show concepts at a relatively low cost, proved to be most supportive to prime-time programmers. Accordingly, networks quickly encouraged more daytime participation from affiliates as well as from independent producers.[5]

The 1950s and 1960s were "the golden years" of television, both in prime-time and in daytime programming. Linda Busby notes the following:

- The number of television stations increased from 107 in 1950 to nearly 900 by 1969.
- Television households grew from 4 million in 1950 to nearly 60 million by 1969.
- The percentage of American homes with television receivers grew from 9% in 1950 to over 95% by 1969.
- From 1950 to 1969, the television industry had grown from a multi-million dollar industry to a multi-billion industry.[6]

It was during this era that much of what we recognize today as traditional daytime television fare was formulated (see Table 2). In addition, programming strategies from independent (or non-affiliated) stations began to emerge as a competitive force (especially during the daytime hours); syndication became a viable alternative to local/network production; and new technologies, such as cable television (CATV) and satellite transmission, became a programming threat to the established authority of broadcast television networks.[7]

THE GOLDEN AGE OF TELEVISION: THE RISE OF INDEPENDENTS

In the early 1950s, the world of television was associated with a very different technological environment than the one in which we live today. For one thing, most TV stations were located on the VHF bandwidth, broadcasting on Channels 2 through 13 only. Also, since most cities had no more than two or three stations, these

▶ **Table 2** Milestones in daytime television programming: 1947–1968

December 25, 1947	NBC introduces a Saturday afternoon children's show entitled, "Puppet Television Theater." This program is later renamed "Howdy Doody,"and is aired five days a week.
August 19, 1950	Saturday morning children's programming is developed by ABC, including two popular shows—"Animal Clinic" and "Acrobat Ranch."
September 18, 1950	Network television presents animated programming with "NBC Comics," a 15-minute weekday show featuring four cartoon heroes: "Danny Match," "Space Barton," "Johnny and Mr. Do-Right," and "Kid Champion."
September 25, 1950	NBC creates a weekday afternoon music/ variety format, with singer Kate Smith as host. CBS quickly follows suit by introducing two hour-long variety shows hosted by Gary Moore and Robert Q. Lewis.
November 1950	WPTZ in Philadelphia introduces a morning show, "Three to Get Ready," with former radio disk jockey Ernie Kovacs. This crazy format is surprisingly successful in its 7:00–9:00 A.M. time slot, and serves as an inspiration for NBC's "Today" show.
September 1951	CBS assumes a leadership role in TV soap operas by introducing two 15-minute programs, "Search for Tomorrow" and "Love of Life."
1950-1951	CBS inaugurates a daytime game show called "Strike It Rich." The program becomes popular enough to be added to the network's prime-time schedule as well.
June 30, 1952	The popular radio soap, "Guiding Light," moves to television.
February 22, 1954	ABC enters the early morning competition with "The Breakfast Club," a successful radio show moved to television. However, despite its congenial host, Don McNeil, the program does not do as well as hoped.
March 1954	CBS premieres "The Morning Show" as network competition for NBC's "Today." "The Morning Show" uses veteran journalist Charles Collingwood as its news anchor and features Walter Cronkite as the program's host.
March 1954	Pat Weaver, the creator of "Today"(NBC), continues his program development success with "Home," a late morning domestically-oriented show. Arlene Francis and Hugh Downs serve as co-hosts.
December 10, 1955	"Mighty Mouse Playhouse," the first Saturday morning cartoon series, is aired on CBS.
1955-1956	Walt Disney combines with NBC to create a late afternoon weekday favorite, "The Mickey Mouse Club."
January 3, 1956	NBC revamps an old MBS (Mutual Broadcasting Service) show, "Queen for a Day" and broadcasts it weekday afternoons.
April 2, 1956	CBS continues its soap opera development by introducing two half-hour shows, "As the World Turns" and "The Edge of Night."

▶ **Table 2** (continued)

November 26, 1956	NBC debuts another game show, "The Price is Right" with Bill Cullen as host.
April 8, 1957	CBS abandons its morning show concept and decides, instead, to present a country-western variety program with Jimmy Dean as host.
August 5, 1957	Philadelphia teen show, "American Bandstand," hits network television as part of ABC's weekday schedule. The local host, disk jockey Dick Clark, stays with the show.
September 30, 1957	A popular night-time show, "Do You Trust Your Wife?," moves to daytime on ABC. The title is later changed to "Who Do You Trust?" and it is hosted by Johnny Carson (with Ed McMahon).
January 6, 1958	CBS drops "Strike It Rich" from its daytime schedule and replaces it with a new quiz show, "Dotto."
August 18, 1958	Facing a potential quiz show scandal, CBS abruptly cancels "Dotto" and replaces it with a new game show, "Top Dollar."
October 13, 1958	ABC introduces *reality programming* to daytime with its new show, "Day In Court."
1959-1960	Syndication comes to TV Programming.
January 3, 1960	CBS Experiments with a new sports concept of featuring lesser known sporting events, such as ice skating and stock car racing on a weekly basis. The show is named, "Sunday Sports Specticular."
December 11, 1961	Local celebrity and former band singer Mike Douglas hosts a 90-minute talk show in Cleveland. By October 1963, the program is syndicated nationally.
December 30, 1963	Monty Hall becomes the most well-known TV huckster on the new game show "Let's Make a Deal."
October 17, 1966	NBC combines celebrities and the popular game of tic tac toe to introduce "The Hollywood Squares." The show becomes an instant hit.
March 4, 1968	ABC inaugurates a morning talk show, featuring comedy writer Dick Cavett. The program seems ill-suited for daytime, however, and is later moved to late night television.

outlets were most likely affiliated with either CBS, NBC, ABC or DuMont.[8] Thus, programming and channel selection were fairly uniform in all major areas of the country.

Things began to change, however, as the demand for more stations grew. The VHF bandwidth soon became insufficient to meet increasing needs, leaving UHF as the only available outlet for possible station expansion. UHF was a technological boon to broadcasting, because it enlarged the potential channel capacity from 12 to 80 stations. On the other hand, programmers soon discovered that mere channel expansion was hardly enough to ensure economic success. This was due to several factors:

1. *Most television receivers were unable to pick up the UHF signal.* Originally, TV sets were designed to receive VHF signals only. To receive UHF stations, it was necessary to purchase a $20 attachment and a special antenna for signal conversion.[9]

2. *The signals received from early UHF stations were less powerful and somewhat less stable than those from VHF stations.* This was because the VHF signal was intrinsically stronger than the UHF bandwidth. In the early 1960s, television technology had not advanced enough to compensate for UHF signal stability variations.

3. *Most television viewers were already in the habit of watching programs on the established VHF stations.* Viewers had become comfortable seeing certain performers, newscasters, and talk show personalities. They were unlikely to switch unless the alternatives were clearly more attractive.

4. *Since most VHF stations were either network-affiliated or owned-and-operated, their programming budgets were much more liberal than those of the newly-formed independents.* As a result, TV shows on VHF were more glamorous and technically slick, utilizing many expensive on-location sets as well as nationally-recognized celebrities (some of whom they had helped to create). UHF, with limited monies and an unestablished track record, could not begin to compete at the same level.

Despite their rather disadvantaged position, however, several independent UHF stations in the early 1960s decided to confront the competition. They succeeded admirably in providing popular programming, even with seemingly overwhelming odds against them. Usually, the growth of independent programming followed a pattern. A description of the process follows. The station manager first determined the "programming holes" in the market from a survey (conducted by an independent research company). In most cases, the results would show that the viewers in the market wanted some other type of broadcast fare, that is, live local sports coverage, and so forth, that was not provided. The independent programmer then set about to fulfill these needs (usually during prime time). To supplement the station's income in other dayparts, TV executives relied heavily on live productions, off-network shows, old westerns, cartoons, and movies (see Figure 1).[10]

Despite the creativity of its programmers, however, most independent stations in the early and mid-60s were seen as the "poor relations" of television. After all,

MORNING

43	6:00	6:30	7:00	7:30	8:00	8:30	9:00	9:30	10:00	10:30	11:00	11:30
	Off-air						Little Rascals		Jack Lalanne	Movie: Dialing for Dollars		

AFTERNOON

43	12:00	12:30	1:00	1:30	2:00	2:30	3:00	3:30	4:00	4:30	5:00	5:30
	Movie, cont.		The Dorothy Nelson Show			Sky King	Sergeant Preston	Bugs Bunny		Rocky and Bullwinkle Show		Popeye

EVENING

43	6:00	6:30	7:00	7:30	8:00	8:30	9:00	9:30	10:00	10:30	11:00	11:30
	Burns & Allen	I Love Lucy	Super-Man	Movie/Sporting Event (In Season)					Lynchville Talks Sports		Off-air	

▶ *Figure 1 A typical independent programmer's day in 1964.*

many of them were UHF-based; as such, they were often labeled as "unprofitable, unsaleable, and subject to unlimited losses."[11]

Fortunately, the plight of the independent programmers began to change drastically in 1970. That year, the Federal Communications Commission (FCC) mandated a Prime Time Access Rule, which 1) limited network prime-time broadcasts to 3 hours each night,[12] and 2) prohibited network affiliates in the top 50 markets from running off-network programming. With these new incentives, syndication companies proliferated to provide competitive first-run shows, franchised programs, co-op productions, and, of course, the ever-increasing numbers of off-network TV fare. Richard Block notes the following in his brief history of syndication:

> . . . in the market where the indies were operating, including the UHFs, network access was a disaster for affiliates. Few seriously considered those dismal little UHF peanut whistles could get sizable audiences programming the "Lucy's," "Adam 12's," "Mary Tyler Moore's" and then a true breakthrough show with "Mash" against the marginal first run fare being offered by the affiliates.
>
> The trends soon made the affils think twice before programming just anything, and they started to run more viewable fare such as "The Gong Show," "Name That Tune" and "Animal Kingdom." Wags called it G & G for gread and gorillas, but the syndicators were now firmly into increased new production.[13]

The late 1960s and early 1970s also signaled changes in television technology, with the introduction of CATV, satellite transmission, and home video recorders. These electronic innovations, along with other economic developments, would serve to change the world of television forever.

NEW TECHNOLOGIES AND TV PROGRAMMING

As already mentioned, the late 1960s and early 1970s are often characterized as a period of great technological and programmatic change in the television industry. To better explain this phenomenon, several historical developments are worth noting:

- In 1967, the Carnegie Commission was appointed to consider the feasibility of a public broadcasting network. By 1969, the Corporation for Public Broadcasting was inaugurated with two production centers: NET (National Educational Television) and PBS (Public Broadcasting Service).
- In the early 1970s, CATV was no longer seen merely as a technological means for wiring remote areas of the nation. Instead, the concept of *narrowcasting*, that is, programming for small defined audiences, took hold, with such channels as the Cable News Network (CNN) in 1972, and Home Box Office (HBO) in 1975.
- Cable's popularity stimulated resource development in commercial satellite transmission, e.g., HBO's satellite-delivered pay-TV (1975). This interest later multiplied into expensive, multifaceted Direct Broadcast Satellites (DBS).
- Broadcast-quality equipment continued to become better as well as less expensive, allowing most stations to produce higher-quality, more localized programming for their viewers.

While the new technologies continued to develop, TV viewers also saw changes in production techniques, program marketing strategies, and syndicated show availabilities, as well as increases in station acquisitions throughout the country. In short, by the late 1970s, most ADI markets had at least four or five TV stations within their coverage areas. Even more often, viewers in these markets had many more channels to choose from via cable or satellite. As a result, network programming popularity began to diminish—more stations had access to a greater portion of audience shares. Accordingly, programming strategies began to diversify to fit specific viewer needs and desires. To further illustrate this point, the following station profiles describe the experiences of three large-market, independent program directors. These examples make it easy to visualize the wide range of strategies available for competing in the TV demographic marketplace:

Station A   This television market includes four independents, as well as the three network affiliates and a strong public station, all competing for a share of the audience. Station A is the "newest kid on the block," having been introduced only seven months ago. There was no question, when Station A began broadcasting, that an aggressive policy was in order to compete favorably with the other stations in the area. Station A's major move was to sign secondary affiliate contracts with all three commercial networks: ABC, CBS, and NBC. This meant that Station A had the option to run network programming that had been refused by the affiliated stations in the coverage area.

Another way of distinguishing Station A from the other competition was to form a partnership with a local FM station. This way, both stations were able to promote each other through shared contests and promotions.

Finally, Station A felt a need to have a special set of personalities associated with it. Since the station scheduled no news in its programming lineup, a goodwill spokesperson campaign was launched so that Station A would have identifiable personnel visiting schools, hospitals, nursing homes, etc. Using network-quality programming, local radio station promotion, and friendly public relations, Station A made a quick impression in its large metropolitan area, and was lauded for its creativity in a crowded television market.

Station B   Station B is one of four independent stations in a seven-station market. As such, it must not only provide alternative programming to the networks, but it must also counterprogram three other indies.

Station B formerly spent most of its time analyzing the network schedule, and then guessing at which series would last long enough to go into syndication. However, with the networks cancelling so many shows after a short term, this task became dubious at best.

Consequently, Station B has adopted a rather unique strategy of analyzing the most popular network programming in its market area, and then approaching producers of such shows after they have been cancelled to see if they will go on syndicating to individual markets. Richard Block notes the following.

> The mid-1980s have stimulated new life in syndication beyond variety shows
> with drama, as with "Fame," and sitcoms as with "Too Close For Comfort," both
> hardly missing a beat after network cancellation.[14]

Sometimes the strategy has worked, and sometimes it has been less than successful. However, with triumphs such as the past hit, "Fame," Station B is confident that it can maintain a high quality of programming, yet continue to distinguish itself from other stations in the market.

Station C Station C has devoted itself to local programming. In fact, more than 11 hours per week (excluding news) of locally produced material is available in both the prime- and nonprime-time day parts in this large metropolitan market. For example, talk/variety shows provide a good alternative to other programming scheduled in the late afternoon, because these shows appeal to a wide audience and are likely to sustain themselves for many years.

Personnel at Station C know that every programmer's wish list is somewhat different, depending on what has worked in the past and what is working in the present. But they firmly believe that a station's true success will rest with its highly identifiable reputation as a local programmer, not with its ability to choose popular syndicated programming.

As one can see from the brief examples listed above, the creative strategies used to present quality local programming are many and varied, due in part to the flexibility of the daytime schedule. Daytime television plays a major role in a station's overall prosperity because it can 1) help to project a strong community image; 2) provide needed revenue for prime-time purchases; 3) serve as a small-scale testing ground for personalities and/or show concepts intended for future use in a later time slot; and 4) appeal to audiences who are not adequately being served elsewhere.

The remainder of this text will address specific daytime programming areas—their costs, their appeal, and their demographic potential. With this information (and a considerable dose of creativity), a local TV station can set itself apart from the rest of the competition.

Notes

1. Harry Castleman and Walter J. Podrazik, *Watching TV: Four Decades of American Television* (New York: McGraw-Hill Co., 1982), p. 27.
2. Linda J. Busby, *Mass Communication in a New Age: A Media Survey* (Glenview, IL: Scott, Foresman and Company, 1988), p. 271.
3. In 1952, ABC aired 20 prime-time hours each week; CBS aired 24.5 hours; DuMont aired 18 hours; and NBC aired 25 hours.
4. Castleman and Podrazik, p. 69.
5. *Daytime programming* is defined as the television fare aired from 7:00 A.M. to 6:00 P.M. (EST), Mondays through Fridays as well as 7:00 A.M. to 4:00 P.M. (EST), Saturdays and Sundays. This definition combines several time slots described in Howard and Kievman's *Radio and TV Programming* (Columbus, OH: Grid Publishing, Inc., 1983), notably *early morning, daytime, late afternoon fringe,* and *weekend daytime.*
6. Busby, p. 273.
7. It was also during this era that DuMont ceased to exist as a network. From 1956 to 1988, ABC, CBS, and NBC were the only commercial networks on the air. In 1985, Fox emerged as a potential fourth network, but it was not officially recognized as such until 1990.

8. Actually, CBS and NBC usually garnered affiliateships in all cities with television stations. If there was a third facility, ABC and DuMont often competed for it. This explains the early second-class status of these two networks and the ultimate demise of DuMont in 1955. It also accounts for the scarcity of independent stations during this era.

9. This began to change later, when the FCC ruled in its All-Channel Receivers Act that all TV sets sold after May 1964 were required to pick up all channels—on both VHF and UHF. But the "VHF only" condition remained much later than 1964, since most people did not rush out to buy a new television set, especially if UHF conversion meant only one additional station in their community.

10. These program offerings represented the first wave of syndication. They usually came from ZIV, a Cincinnati-based producer/distributor, which gathered station business, then sold advertising time to regional sponsors.

11. Richard C. Block, "History of Syndie, A Short Look Back," *Variety* (February 17, 1988), p. 77.

12. The actual rule allows for network programming from 8:00-11:00 P.M. (EST), Mondays through Saturdays; and from 7:00-11:00 P.M. (EST), Sundays.

13. Block, p. 77.

14. *Ibid.*

2

▼ Talk Shows/News Magazines
▼
▼
▼
▼

Despite the fact that a lucrative advertising revenue can be obtained from appealing to any demographic viewing group, many broadcast executives still prefer to target most of their programming to 18 to 49 year-old women. The reasons are clear:

1. Research shows that the greatest percentage of habitual TV viewers fall into this age and gender category.
2. Most women still take the responsibility for buying household items.

Thus, with the exception of Saturday mornings and weekend afternoons (the primary viewing hours for children and adult males), daytime strategists often gear their major programming efforts towards reaching adult women. This does not mean that other demographic groups are unimportant during weekday mornings and afternoons. In fact, the subsequent chapters of this book would deny such an assertion. It is, however, important to recognize that a station (or network) must prioritize its audience subsets before deciding which shows to air.

With this in mind, one of the most profitable daytime programming categories for adult women throughout television history has been the talk show/news magazine. Several variations may exist within this format, namely, early morning shows, talk/variety programs, infotainment, and so forth; but all must maintain the objective of informing the audience, while giving them a healthy dose of entertainment.

EARLY MORNING SHOWS

As mentioned in Chapter 1, the concept for a network early morning show was derived, in part, from a creative local TV program, named "Three to Get Ready," aired by an NBC station in Philadelphia in 1950. Hosted by former radio disc jockey Ernie Kovacs, the weekday morning broadcast included a brief look at current news and weather, along with some lighthearted fun. While Kovacs's personality may have been better suited for prime-time television, his early morning show nonetheless proved that a substantial (and desirable) viewing audience could be found on weekdays from 7:00–9:00 A.M.[1]

The NBC program director at that time, Sylvester "Pat" Weaver, took note of "Three's" apparent popularity and developed a similar format for network broadcast. His original title proposal for the show had been "Rise and Shine." The actual program was later renamed "Today" and budgeted by NBC at $40,000 per week.

"Today" was hosted by a Chicago personality named Dave Garroway, who had become recognized for his informal, laid-back manner. Garroway's demeanor and Weaver's concept seemed unbeatable to NBC programming executives. In fact, the network was so excited by the new concept of *morning television*, it spent large amounts of money to promote the show's 1952 premiere as a "revolution" in broadcasting.

Unfortunately, the initial critiques of "Today" were not as positive as NBC had hoped. Castelman and Podrazik describe the first broadcast as follows:

> . . . the first *TODAY* broadcast appeared as an almost meaningless hodge-podge. The cast and crew were squeezed into a tiny street-front New York City studio that had originally been a public display showroom for RCA TV sets, and viewers could see: three teletype machines, weather maps, wirephoto displays, clocks set to the times of various world cities, record players, newspapers, the crowd outside the studio, and, oh yes, the show's regular cast of Dave Garroway, Jim Fleming, and Jack Lescoulie. Throughout the program there were frequent cuts to live reports from the Pentagon, Grand Central Station, and the corner of Michigan Avenue in Chicago, as well as live phone reports describing the weather in London, England and Frankfurt, Germany. Viewers were bombarded [sic] with data and the first reactions to *TODAY* were confusion and indifference. *TODAY* did not seem to have any point other than to show off fancy gadgets.[2]

Despite the damaging reviews, however, Weaver and his "Today" producers continued to work with the basic premise of the show. They readjusted the balance between technological gadgetry and Dave Garroway's quiet personality. By May 1953, the program had begun to reflect its true potential.

"Today's" success soon became a breeding ground for many wishful contenders, both locally and nationally. Ernie Kovacs, for example, was bumped from his old timeslot at NBC's WPTZ. He later moved to New York to host a local early morning show on the CBS owned-and-operated station there. As for network competition, CBS premiered "The Morning Show" in 1954; and ABC countered with a televised version of the popular radio program, "The Breakfast Club" during that same year. Both productions paled in comparison to "Today," however, and were ultimately cancelled. CBS later decided to go after a younger demographic in its early morning time slot and created "Captain Kangaroo" (which became a children's icon for many years until its cancellation). ABC, on the other hand, urged its affiliates to develop local programming for several seasons before initiating its next network show, "A.M. America"(later re-vamped and renamed "Good Morning America") in 1975.

Meanwhile, local stations throughout the country developed morning shows similar to the "Today" show format. In these settings, a popular host or hostess would introduce various segments on subjects such as cooking, household problems, community affairs, and local entertainment. News and weather reports usually supplemented these featured spots, creating a balanced mixture of information and entertainment to start the day. Often, the morning show personality was strongly identified with the local station; thus, a popular program in this timeslot could benefit the TV station beyond generating revenue for the early daytime hours.

In the 1970s, CBS continued to devote its weekday morning resources to children's programming (although it also renewed its commitment to AM informational broadcasting through "The CBS Morning News"). Conversely, ABC decided to confront NBC's "Today" show head-on with "Good Morning America."

ABC's bold move had not been ill-conceived. While "Today"was still a successful programming vehicle at this time, it had not been without its problems, including some tumultuous personnel changes—both behind the camera and in front of it. Also, the network decisionmakers at NBC had become a bit too complacent about their morning show's popularity. As a result, it was not surprising that ABC, "lean and mean," would mount a strong, competitive campaign against its early morning rival.

"Good Morning America," with David Hartman and Nancy Dussault, appealed to a younger audience than NBC's "Today. " The show's pace was quicker, the graphics were more contemporary, and the hosts were extremely likeable. In addition, the feature segments were generally more state of the art, emphasizing personalities and topics that would appeal to a mid-30s or younger demographic.

Before long, "Good Morning America's" ratings began to threaten the executives at NBC. By 1978, ABC had actually launched a successful programmatic attack against "Today's" early morning daypart—a coup which had never taken place before.

Throughout the 1980s and into the 1990s, "Good Morning America" and the "Today" show have continued to vie for ratings' superiority in the 7:00–9:00 A.M. time slot. Celebrity dramas, such as the Jane Pauley-Deborah Norville popularity dispute, David Hartman's alleged dislike of Joan Lunden, or Bryant Gumbel's notorious memo concerning Willard Scott, serve only to fuel the continuous fire of competition between ABC and NBC.

Unfortunately, however, "CBS This Morning" is yet another ill-fated attempt by this network to compete with "Today" and "Good Morning America" for the early morning adult demographic. Shortly after the show was first aired (under its original title, the "CBS Morning Program"), many critics felt strongly that CBS should have never abandoned its perennial childhood favorite, "Captain Kangaroo" (and its lucrative revenues). To date, this sentiment has not changed. Thus, the early morning programming hours are still dominated by ABC and NBC. Many CBS-affiliates and independent TV stations use this time block to appeal to alternative (or underserved) audiences instead.

TALK / VARIETY SHOWS

All television talk shows, whether entertainment-oriented or informational, hold one identifiable characteristic in common: the importance of a strong interview. Thus, the secret of a specific program's success can be traced to the host, who should be very adept at both talking and listening. TV scriptwriting authors Edgar E. Willis and Camille D'Arienzo describe the talents of a good interviewer in the following way:

> They are listeners, reacting to bits of information with a good sense of the mood
> and meaning of what is said or left unsaid, by their guests. Successful interview-

ers put their guests at ease in order to draw from them information they may not have shared elsewhere. Their attentiveness enables them to respond to the unexpected confidence, to encourage the disclosure, and to remain flexible enough to lay aside the prepared questions, should more interesting information begin to surface. Yet good interviewers must remain firm enough to redirect the evasive guest to the body of information both agreed they would discuss. Interviewers must be aware of their responsibility to ask the questions members of the audience would raise, were they in a position to do so. If they tend to be thorough researchers, they must take special care not to overlook the obvious, yet often necessary questions important to their listeners.[3]

Like most program genres discussed in this book, the talk show format began in radio, and later moved to television because of its relatively low cost and high appeal.[4] The transition from radio to TV was not an easy one, however, because the visual dimension of the new medium was not well-suited to the prewritten, prerehearsed extemporaneous interviewing done by radio personalities in the past. Consequently, a new, impromptu method of presentation was needed. For several years, the networks experimented with various talk show forms to enhance the genre's popularity on the screen.

The first program of this type was NBC's "Meet the Press" (1947), in which a moderated panel of journalists often interviewed a public figure about some current political event or economic trend. Later, in 1954, CBS introduced a similar show, "Face the Nation," and both broadcasts enjoyed a relatively healthy audience response.

"Meet the Press" and "Face the Nation" were aired only once a week, however, usually on Sunday mornings. Apparently, the rather formalized public affairs conference did not play well on Monday through Friday. Still, talk show television had shown some potential: it simply needed several modifications to be successful on a weekday schedule.

NBC programmer Pat Weaver was the first network executive to appreciate the varied possibilities of TV interviewing. After developing the "Today" show, Weaver set about to create another casual, entertainment-oriented talk program for late morning weekday viewing. The results of his effort premiered in March 1954, with "Home"—a clever blend of the "Today" show and the traditional interview format.

Drawing from "Today's" audience flow, "Home" was geared predominantly toward housewives, but offered them more glamorous fare than the typical morning offering of recipes, household advice, and tips on child-rearing. The show won immediate applause from viewers throughout the country, and continued to do well until 1958, when CBS's growing soap opera schedule (by now increased to "Love of Life," "Search for Tomorrow," "The Guiding Light," "As the World Turns," and "Edge of Night") took over.

For the next several years, most late morning and early afternoon talk shows were locally produced. As such, they were limited greatly in size and scope. Despite their obvious flaws, however, local talk shows became quite popular among their target audience of adult women. Unlike the sophisticated banter provided by well-known celebrities like Arlene Francis and Hugh Downs, these programs featured *local* personalities discussing *local* events with which viewers could identify. Obviously, TV stations were not able to accrue many national sponsors for their low-budget productions; but small-town advertisers loved the time slot and format. More-over, local television executives saw talk programming as a sound investment. It was cost-efficient as well as a strong image-maker for the station.

One of the most successful local talk programs during this era was "The Mike Douglas Show," a weekday entertainment vehicle produced by station KYW in Cleveland, Ohio. Westinghouse (later known as "Group W"), the owners of KYW, had decided to hire Douglas as a local talk show personality because of his singing talents,[5] his conversational abilities, and his sincere manner. Their investment in the program soon paid off. In 1963, just two years after "The Mike Douglas Show" debuted, it was syndicated and distributed nationwide.

Group W's success with Mike Douglas opened the syndication doors for more daytime talk/variety shows in the 1960s and 1970s. Programs with personalities such as Merv Griffin, Dinah Shore, Virginia Graham, Phil Donahue, Dick Cavett, John Davidson, and David Letterman battled heavily for ratings victory at this time. In the aftermath, only "The Merv Griffin Show" and "Donahue" survived long enough to move into the 80s. The other programs were either cancelled or reformulated and moved into a later timeslot.

For some years, Phil Donahue dominated the talk show airwaves, overpowering Merv Griffin in the Nielsens, and overshadowing other similar broadcast fare, like "Hour Magazine" and "Sally Jessy Raphael." "Donahue's" success was attributed to his salient mixture of entertainment and social issues. In fact, many of his viewers often referred to him as "an intelligent person's talk show host." However, in the mid-80's, several first-run syndicated programs challenged "Donahue's" domain. Perhaps the biggest surprise in this arena was "The Oprah Winfrey Show," featuring a Chicago-based television personality who added high emotional fervor to her talk show topics.

Audiences loved Oprah and the way she interviewed her guests. Soon, programs such as "Geraldo" and "The Morton Downey Show" followed syndication suit, emphasizing emotion over intellect. These programs were subsequently referred to as *trash TV* or *tabloid talk*. Most media critics condemned this new variation on the interview format, saying it was unprofessional, exploitative, and sleazy. However, viewers seemed to applaud it; and as the ratings climbed, so did the tantalizing topics for TV discussion. Geraldo Rivera continued to brawl with his audience both physically and verbally, ultimately having his nose broken; Sally Jessy Raphael's shows added a new dimension to the word sensationalism; and Oprah Winfrey dramatically revealed more and more details of her personal life to the camera. Phil Donahue even wore a dress during one of his shows to gain more media attention.

At this writing, tabloid talk programming seems to be an important daytime component in the schedules of most local stations. This phenomenon is likely to continue, at least for the next several years, due to advertising support and viewer popularity.[6] Even some of the former critics of tabloid talk have changed their perspective considerably. Recently, one author explained this type of programming in the following way:

> These tabloid talk shows can be great fun. Sometimes they are relevant. And sometimes they are even—brace yourselves—good journalism. With the correct topic and motivated guests, a talk master can reveal the dimensions of significant human issues with a clarity and reality beyond the grasp of print journalists.

Talk Shows—1991 Availabilities
Current First-Run Programs

Program	Distributor
Donahue	Syndicated (Multimedia)
Dr. Edell's Medical Journal	Syndicated (HMS Ent.)
Face the Nation	Network (CBS)
Fight Back with David Horowitz	Syndicated (Western Intl.)
Everyday with Joan Lunden	Syndicated (Michael Kraus Prods.)
Geraldo	Syndicated (Paramount/Tribune Ent.)
House Party	Syndicated (Group W/w NBC station group)
Just Between Us	Syndicated (GTG)
Kelley and Gail Show	Syndicated (Tribune Ent.)
Live with Regis and Kathie Lee	Syndicated (Buena Vista)
Meet the Press	Network (NBC)
My Talk Show	Syndicated (MCA)
Not for Men Only	Syndicated (Viacom)
Sally Jessy Raphael	Syndicated (Multimedia)
Studio 33—Hollywood	Syndicated (MCA)
The Joan Rivers Show	Syndicated (Paramount)
The Oprah Winfrey show	Syndicated (King World)
This Week with David Brinkley	Network (ABC)
Voices of America	Syndicated (Warner Bros.)

Anyone who caught the recent programs in which Oprah extracted from her guests gripping personal dramas of racism and aging and infidelity had a true learning experience. She revealed universal truths and lessons of relevance and immediacy to her audience.[7]

Talk shows have evolved into a very different genre when compared to Pat Weaver's original concept in 1954. In addition, their seeming success has paved the way for an entirely new genre of news and information presentation in the 1990s—*infotainment.*

NEWS MAGAZINES AND INFOTAINMENT

In 1988, producers at Twentieth Century Fox TV produced and distributed a news magazine that would come to be called *tabloid journalism.* "A Current Affair," hosted by Maury Povich, focused on soft news topics of human interest, and presented them in a fast-paced, technologically slick manner. While many local station owners were dubious about the show's potential, several programmers decided to take a chance and aired it in the late afternoon fringe period (after the network soap operas and opposite "Donahue" and "The Oprah Winfrey Show" in most markets).

Shortly after its rather inauspicious beginning, "A Current Affair" became one of the most talked-about shows in the 1988 season. Viewers loved it, critics hated it, advertisers supported it, and rival production companies rushed to copy it. Predict-

News Magazines—1991 Availabilities
Current First-Run Programs

Program	Distributor
A Current Affair	20th Century Fox
Business this Morning	Viacom
Byron Allen	Genesis
Celebrity Update	GTG
Current Affair Extra	20th Century Fox
Entertainment This Week	Paramount
Entertainment Tonight	Paramount
First Business	Biz Net
Hard Copy	Paramount
Inside Edition	King World
Inside Report	MCA
Inside Video This Week	MG/Perin
Personalities	20th Century Fox
Preview	TPE
Private Affairs	Multimedia
TV Personals	LBS
USA Today	GTG

ably, within a year, several new tabloid magazines flooded the marketplace, with shows such as "USA Today: The Television Show,"[8] "Tabloid," "Exclusive," "Inside Edition," and "Hard Copy. " In addition, some of the more established syndicated magazine programs like "Entertainment Tonight" and "Evening Magazine" revitalized their formats, giving them a faster pace and more provocative story selection.[9]

The entertainment-oriented news magazine (or *infotainment*, as *Variety* has dubbed the new genre) quickly became a gold mine for several delighted production companies. For example, writer John Dempsey reported the following in 1989:

> ...if the seven Fox stations that strip "A Current Affair" ponied up [their] license fees commensurate with the ratings the show deliver[ed], Fox would rake in $500,000 a week. Fox also pocket[ed] the additional $350,000 a week that "Affair" generate[d] from the two national 30-second spots in each half-hour, giving the company a total gross of $850,000 a week. That's a staggering figure when balanced against the production cost of the show, which one insider says [came] to about $260,000 a week.[10]

Despite the rapid financial gains in the past two years, however, several potential infotainment problems still loom in the background, including the following:

- Surveys reveal that tabloid news is often seen as equivalent to traditional journalistic programming by many viewers. More significantly, a large percentage of children and young adults seem unable to distinguish between the two genres. Combined with the alarmingly high number of functional illiterates found in this country, tabloid news shows are said to be contributing to the intellectual delinquency of America's youth.

- The growing competition for ratings supremacy among tabloid news magazines has created a fever among producers for overly sensationalistic, exploitative topics to be aired. As a result, some advertisers are beginning to cool in their ardor for commercial time on such shows. In March 1990, for example, 24 advertisers (who collectively spend more than $2 billion each year on national spots) chose to boycott "A Current Affair," "Inside Edition" and "Hard Copy."[11] The trend is quite likely to continue if the topic choices do not change.

- The meteoric rise of infotainment may be short-lived. Considering the large amount of format replication which is presently going on (either in actual production or at the developmental programming stage), a saturation point is inevitable. Accordingly, the viewing audience will soon tire of the genre's overexposure and move on to other daytime fare.

Thus, tabloid news shows are not as firmly established in TV programming schedules as are other entertainment forms. However, their ultimate longevity or obscurity will be an interesting observational adventure in the new few years.

Notes

1. The morning hours here are based on Eastern Standard Time (EST).
2. Castleman and Podrazik, p. 70.
3. Edgar E. Willis and Camille D'Arienzo, *Writing Scripts for Television, Radio and Film* (New York: Holt, Rinehart and Winston, 1981), pp. 141-142.
4. Talk shows are extremely inexpensive programs to produce (second only to game shows in cost). The primary investment for such programming is the host's salary, as well as whatever expenses (hotel, travel, meals, etc.) the guests may accrue. Since most celebrity guests are paid only the minimum in union scale for their appearances, the latter costs are usually quite small.
5. Mike Douglas had previously sung with Kay Kyser during the big band era.
6. Many advertisers strongly support shock-oriented topics on shows such as "Geraldo," "The Oprah Winfrey Show,"and "Donahue." However, some of the more traditional, "blue chip" sponsors (like Proctor & Gamble) are beginning to shy away from certain topics on certain programs (see footnote 11). If this trend continues and broadens, the amount of sensationalism on talk shows will surely decline sharply.
7. Van Gordon Sauter, "In Defense of Tabloid TV," *TV Guide* (August 5, 1989), p. 4.
8. This program was later renamed "USA Today on TV."
9. While "Entertainment Tonight" and "Evening Magazine" are aired primarily during the prime-time access period, they have been included in this section as examples of more traditional news magazines which have changed as a result of the popularity of "A Current Affair" and its imitators.
10. John Dempsey, "More Mags Will Fly in Fall; Too Much of a Bad Thing?," *Variety* (April 4, 1989), p. 79. Dempsey went on to say that the $260,000 figure would change somewhat, since it did not include the new $1.8 million-a-year contract Maury Povich signed in 1989.
11. John Dempsey, "24 Advertisers Boycott 'Affair,' 'Edition,' 'Copy,'" *Variety* (March 21, 1990), p. 1. The boycotting advertisers (along with their total national/regional/local

spot expenditures for 1989) were listed accordingly: General Motors ($350 million); Pepsico ($265 million); Proctor & Gamble ($220 million); Ford ($180 million); General Foods/Kraft ($179 million); Bell Telephone's regional companies ($155 million); Walt Disney ($122 million); Sears ($108 million); Honda ($68 million); Coca-Cola ($68 million); Chevron ($49 million); Lever ($48 million); Time Warner ($46 million); American Express ($43 million); AT&T ($40 million); Kellogg ($39 million); Mars ($37 million); Clorox ($33 million); Heinz ($30 million); Citicorps ($25 million); Winn-Dixie ($21 million); Hormel ($21 million); Eckerd ($17 million); and Levitz ($17 million).

3

▼ Game Shows

According to most media sources, an estimated 100 million people regularly watch game shows in the United States each week. Viewers come from all demographic groups—kids, teens, adults, senior citizens, rich and poor, highly-educated and functionally illiterate[1]—and network executives love them all.

Television program directors are particularly attracted to game/quiz shows because they are so inexpensive to produce, yet yield a high percentage of profit. Most games and quizzes cost an average of $20,000 per show (or $100,000 per week) to produce. This figure compares most favorably with every other program genre available. For example, a half-hour daytime soap usually costs about $60,000 per show (or $300,000 per week); a Saturday morning children's cartoon might cost $150,000 for 60 minutes; and an average half-hour prime-time drama commonly costs about $475,000.

Since network television began in 1946, approximately 400 game/quiz shows have debuted in America. In fact, a day has never passed since that time without at least one such program on the air.[2] Perhaps this is due to the immediate popularity that question-answer programs enjoyed from the very beginning.

The TV game show genre started as have most television programs—as a revamped version of some popular radio hit. This is because network programmers in the late 1940s hoped that listeners would convert more easily to television if they were able to enjoy the same shows they tuned into on their radio receivers. In the context of game shows, the most successful of these television clones were stunt programs with visual appeal, such as "People Are Funny" and "Truth or Consequences." In each case:

> The format consisted of handsome interchangeable male hosts using the lure of cash or gifts to lead unsuspecting, eager, basically naive average Americans through demeaning yet undeniably funny shenanigans. In these audience participation programs, antics were much more entertaining when seen rather than merely described.[3]

In addition to the "physical" entertainment shows, broadcast executives soon discovered that competitive quiz programs were often extremely popular with many segments of the television audience. The best example of this format was "Uncle Jim's Question Bee," which premiered on radio in 1936, and ultimately became the blueprint for such program descendants as "Information Please," " Pot O' Gold," "Quiz Kids," and "Stop the Music."

Of the programs listed above, however, "Stop the Music" varied most from the formula established by "Uncle Jim's Question Bee. " First of all, winning contestants were often awarded *prizes* instead of small sums of cash (as were given on other quiz shows). In addition, "Stop" appealed as much to the greed of home audiences as it did to that of the studio participants. Producers accomplished this objective by offering monetary incentives for those listening at home. Castleman and Podrazik describe the programming hook as follows:

> Phone numbers from across the nation were selected at random. While host Bert Parks dialed a number, the show's musical regulars began performing a popular song. As soon as the home contestant picked up the phone, Parks would say "Stop the music!" and ask the listener to identify the song that had just stopped playing. If correct, the listener would win a prize and a chance to identify a much more difficult "mystery melody" worth as much as $30,000. Though the odds against being called were astronomical, listeners felt it was wise to tune in and be prepared—just in case.[4]

Indeed, "Stop the Music" and many other early programs were successful TV descendants of radio. However, as time progressed, the visual medium of television proved to be significantly unique enough to generate its own game/quiz show format. "Cash and Carry" was the first such network show specifically designed for this purpose. "Cash " was produced and distributed by the DuMont network (premiering on June 20, 1946) and was a variation of the old "Truth or Consequences" theme—if contestants answered a question correctly, they were given a cash prize; if they answered incorrectly, they were forced to pay the consequences. "Cash and Carry's" initial success further inspired DuMont to create other game shows such as "Play the Game," a charades-type program hosted by New York celebrity Dr. Harvey Zorbaugh.

But DuMont was not alone in its enthusiasm for the game/quiz show genre. By 1949, over 15 games (in both daytime and prime-time) filled all four network schedules. In addition, locally produced quiz shows abounded, since these programs were so popular and economical to air. Needless to say, this game show explosion fueled the desire to improve upon the basic genre format. It also inspired programs that contained the signature of the network and the production company.

The first of these signature programs was "Winner Take All," a production from the now famous team of Mark Goodson and Bill Todman. "Winner" was the first show to feature the *carryover contestant* — someone who could continue to play game after game until he or she lost to a new champion. "Winner" also originated the concept of using a buzzer to determine which contestant could answer the question posed by the show's host. Both innovations were received enthusiastically, and Goodson and Todman went on to produce "Beat the Clock" and "What's My Line?" in the early 1950s, as well as many other contestant games in the decades ahead.

"What's My Line?" was actually remembered as the first "hybrid"—a combination quiz and game show—in television history, because it emphasized celebrity reactions to the questions as well as the questions themselves. Producer Goodson says the following:

"What's My Line?" transformed the classic schoolroom quiz into something else—the game. Instead of doing questions, we turned everything on its ear. We cast a panel of well-known people, but used as the puzzle to be solved a real human being. The panel was challenged to solve a human problem, rather than an academic one. It became a game of real life. It was developed prior to the days of talk shows. So "Line" was more than a game. It was an elegant talk show, *plus* a game.[5]

The notion of celebrities participating in some type of gamesmanship became popular almost immediately. As a result, one of the major characteristics of 1950s game/quiz shows became celebrity appearances. Some of the most notable participants during this era were Jack Paar, Sam Levenson, Henry Morgan, and Johnny Carson. Obviously, these gentlemen went on to other things as their careers progressed. However, they will always be known as "game show alumni."

Another popular trend in the 1950s was *sob programming*. Adapted from radio, these shows (like "Queen for a Day" and "Strike It Rich") featured poor people who appeared on stage to tell their personal traumas. Most needed money for some special reason (such as a required operation, an adoption of a war-torn orphan, or a washing machine for an overgrown family). These stories were then given consideration by the studio audience, voted on (by an applause meter) and awarded distinctions (the highest prize being "queen for a day"). The winner was then given the money or object needed, and (allegedly) lived happily ever after.

The *sob show* syndrome exposed a potential for dramatic flair that few network executives had previously thought possible. Thus, as more producers continued to create more ideas for games, the implicit demand for higher profile programming emerged—with the ultimate result—the quiz show scandals.

They evolved slowly—and innocently—at first. Producer Louis G. Cowan (creator of "Stop the Music") had become interested in updating an old radio quiz program called "Take It or Leave It." The original version of the show had begun with an $8 question and advanced steadily until the winnings totalled $64. Cowan liked the pyramid concept, but he wasn't pleased with the money values. Thus, he started to consider other totals. $640? $6400? Neither seemed to be very impressive. However, $64,000 started to whet his appetite in the areas of both money and weekly drama—"The $64,000 Question" had been born. The format of the show was as follows, according to Jefferson Graham:

[The first show] began with a sixty-four-dollar question, and the questions doubled in value seventeen times, each increasing in difficulty, all the way up to sixty-four thousand dollars. The contestant would answer some easy questions, then climb into an isolation booth at the eight-thousand-dollar level, where he or she pondered an answer for thirty seconds of dramatic background music. Having successfully answered, the contestant would leave and come back the following week, at which time he or she announced the decision: take the money and walk away or risk the next plateau, double or nothing—well, not nothing. You got a Cadillac as a consolation prize. . . [It] was every American folk tale rolled into one—the Horatio Alger story played for big bucks, live, every Tuesday. Study, work hard, and you, too, can become rich and famous.[6]

The show concept ignited audience enthusiasm immediately; and other producers began creating imitations of "64" almost as quickly. In addition, the previously established quiz programs (which had been giving away lesser sums of prize money) were forced to raise their stakes or lose popularity. From this point on, there would be no return to simple, uncomplicated game show structures.

Among "The $64,000 Question's" competitors was the NBC production, "Twenty-One." It was a bit more technically slick and complex than its forerunner; but, for some reason, it lacked the audience loyalty that "The $64,000 Question" seemed to inspire. As a result, the show's creator, Dan Enright, felt severe pressure from his network and "Twenty-One's" advertisers to rig the results.

Initially, the rigging scheme was unsuccessful, since Enright had trouble finding contestants who would go along with his plan. However, after several months of failure, Enright met Charles Van Doren, an associate professor at Columbia University, who wanted the fame and glory (as well as the wealth) of being one of the most visible game show contestants in history. Unfortunately for Van Doren (and, incidentally, for Enright and NBC), his fame and wealth were short-lived. To some media critics, though, Charles Van Doren did achieve the dubious distinction of becoming one of the most *notorious* game show contestants in history.

In 1959, after the quiz show scandals were exposed, Congress passed a law making it illegal to rig any game show. In response to this action, NBC, CBS, and ABC [7] fired and blacklisted many of their personnel. They also established departments of standards and practices to censor informally all game shows as well as other types of programming.

As for Dan Enright, the producer of "Twenty-One," he admits that what he and the network did was unforgivable—and unforgettable:

> We misled the people, we misled the viewers, we betrayed their trust. We did it because we thought it was a form of entertainment. In an effort to provide better entertainment we lost sight of what we were doing. I will never lie again. It's not worth it. I'll simply tell the truth and bear the consequences. [8]

After the quiz show scandals of the late 1950s, network programmers became extremely cautious about any further incidents that might taint their reputations. Thus, the TV games of the sixties were much more restricted than those before them. Question-answer programs during this era included "Jeopardy!," "Password," and the less successful "Video Village." Despite some obvious differences in actual format, [9] the common thread connecting all of these shows was an emphasis on low-money prizes as well as a high level of game purity. In other words, broadcast executives were committed to the idea of keeping the quiz/game genre because of its previous success; however, they wanted to make sure that no one would ever associate them with any impropriety again.

One program—"The Hollywood Squares"—admittedly continued to rig some shows; but this was done openly, [10] as a means of making the celebrity participants look wittier than they might have been without briefing. It also added a dimension that had been introduced on another successful show of the 1950s, "To Tell the Truth. " By giving some of the celebrities some of the answers before the program, contestants not only had to determine the correct answer to the question, they also had to judge the veracity of any given celebrity's comments. The formula also gave celebrity participants more confidence, which, in turn, made them more congenial and entertaining.

Another popular trend in the 1960s was a game show format that was wild, oddball, and virtually without content. The major goal in these programs was to determine how deeply people would embarrass themselves to win a prize. Obviously, network executives had no need to worry about show rigging, since there was no real knowledge required to play the game. And viewing audiences loved the vicarious pleasure of watching persons make fools of themselves. Distinctive programs of this type included "Let's Make a Deal," "The Dating Game," and "The Newlywed Game."

In contrast to the 1960s, TV games in the 1970s became nothing more than revisions of successful shows from earlier years. During this time, *TV Guide* began listing programs such as "The New Price is Right,"[11] "Password Plus,"[12] and "The New Name That Tune."[13] In addition, several shows (like "Tic Tac Dough" and "You Don't Say") were mildly revamped, but they still kept their original titles. In short, rarely more than mere cosmetic changes[14] were applied to these programs since they had already proven themselves to be a safe bet.

Aside from vintage revisions and a continuation of wilder games (like Chuck Barris' "Gong Show"), the main distinction of daytime game programming in the 1970s was the tendency to include sexual innuendo. Shows like "The Match Game" often asked such questions as, "Harvey always thought Melvina's _____ were too large." The direction of the responses was all too obvious; thus, viewers enjoyed the playful manipulation by both celebrities and censors. "Tattletales" went one step further by featuring celebrity couples and their personal secrets in everyday life. Viewers loved to learn that Glenn Ford might leave his dirty socks in the kitchen or that Bill Macy might enjoy eating spaghetti in bed. Incidentally, "Tattletales'" host Bert Convy became a celebrity in his own right at this time. In addition to his daytime stint as game show moderator, he often accepted cameo roles in prime-time television. He also attempted a singing career, which unfortunately, was somewhat short-lived.

Another interesting development in 1970s game shows was the heavy emphasis on gambling. Surprisingly, too, was the fact that this unique approach was received favorably by network censors. In fact, most broadcast executives agreed that, if anything, the implication of luck and chance in programs such as "Gambit," "The Joker's Wild," and "High Rollers," precluded any illusions of show rigging found during the 1950s quiz show scandals. They heartily endorsed the concept.

As a result of such aggressive programming techniques, the mid-70s is often seen as the "golden age" of game show production. In fact, in 1975, more than 26 game shows were featured in syndication and on network television.[15] However, by 1980, viewing audiences had seemingly begun to tire of this type of daytime fare. As a result, several programs were soon cancelled, and few game/quiz proposals were accepted. In fact, only one new game show, "Scrabble," succeeded during this time. In short, traditional show formats were returning to game television; but despite this change, networks were still unwilling to accept most suggestions.

However, this network resistance paved the way for syndicated programming development. By the early 1980s, many new TV stations—both affiliated and independent—had emerged throughout the country. In addition, most stations had begun to operate for 24 hours. Thus, station owners were constantly seeking inexpensive, high-yielding program fare.

Within this context, an innovative producer named Merv Griffin entered the broadcast scene with his new creation, "Wheel of Fortune." "Wheel" became a huge success almost immediately, with profits upward of $110 million annually.[16] It also allowed Griffin to introduce his other pet project, a new version of "Jeopardy! "

Both "Wheel of Fortune" and "Jeopardy!" have since become the signature game programs of the 1980s. They have also inspired similar show development. In fact, at the 1990 annual NATPE[17] convention, Bob Jacquemin (president of Buena Vista TV) commented that the 1990–1991 season was definitely "the year of the gameshow. There [were] at least 25 new gameshows in development for [the] fall."[18] Janeen Bjork (vice president of programming for Seltel) concurred with Jacquemin's observation, and add, the following:

> All of the pilots I'm aware of are cheaply produced. There are no high-budget shows with big production values, or shows that are trying for a decidedly different look. In this marketplace, the trend is toward gameshows that cost only $20,000 in half-hour to produce.[19]

After analyzing these TV production figures for the early1990s, it is clear that "Wheel of Fortune," and "Jeopardy!" have successfully reinstituted the game/quiz show format among viewers, programmers, and advertisers. And their low cost, high yield reputation will certainly secure them a place on station and network schedules in the near and distant future.

GAME SHOW FORMS

As a final note, according to Jefferson Graham, in *Come On Down! The TV Game Show Book*, there are nine basic quiz/game show formats in existence today. Thus, all of the programs listed in this chapter can either be classified as one of these types or as combinations thereof:

- **Question and answer:** the basic quiz show format found in "Uncle Jim's Question Bee," "Sale of the Century," and "Jeopardy!".
- **Words:** the televised crossword puzzle, where contestants fill in letters to form words. Examples of "word" shows include "Wheel of Fortune," "Scrabble," and "Crosswits."
- **Word communication:** the transmission of clues by celebrities to non-famous contestants, allowing them to win a prize if they guess the mystery word or phrase. Examples of this format are programs such as "Password," "$25,000 Pyramid," and "You Don't Say."
- **Puzzle:** the process of answering initial questions correctly as a means of revealing clues to a hidden puzzle underneath the first layer of the gameboard. "Concentration" and "Tic Tac Dough" fit into this category.
- **Panel:** several celebrities (three or four) match wits with a mystery guest who possesses some secret, unusual occupation, and so on. "What's My Line?," "To Tell the Truth" and "I've Got a Secret" are all examples of this quiz show prototype.

Game Shows—1991 Availabilities

Program	Distributor
A Question of Scruples	Syndicated (Worldvision)
Best of Groucho	Syndicated (W.W. Ent.)
Bumper Stumpers	Syndicated (MG/Perin)
Chain Reaction	Syndicated (Bob Stewart Prod.)
Challengers	Syndicated (Buena Vista)
College Madhouse	Syndicated (Warner Bros.)
Concentration	Network (NBC)
Face the Music	Syndicated (Sandy Frank Prod.)
Family Feud	Syndicated (Lexington)/Network(CBS)
High Rollers	Syndicated (Century Towers Prod.)
Hollywood Squares	Syndicated (Orion)
Jackpot	Syndicated (Palladium)
Jeopardy!	Syndicated (King World)
Joker's Wild	Syndicated (Orbis)
Love Connection	Syndicated (Warner Bros.)
Monopoly	Syndicated (King World)
Name That Tune	Syndicated (Orion)
Press Your Luck	Syndicated (Carruthers Company)
Price is Right	Network (CBS)
Puzzle Game	Syndicated (Tribune Ent.)
Quiz Kids Challenge	Syndicated (Barris)
Rodeo Drive	Syndicated (Jay Wolpert Prod.)
Remote Control	Syndicated (Viacom)
Scrabble	Network (NBC)
Supermarket Sweep	Syndicated (Al Howard Prod.)
Talkabout	Syndicated (Taffner)
3rd Degree	Syndicated (Warner Bros.)
Tic Tac Dough	Syndicated (ITC)
Trump Card	Syndicated (Warner Bros.)
$25,00 Pyramid	Syndicated (Bob Stewart Prod.)
Wheel of Fortune	Syndicated (King World)/ Network (CBS)
Win, Lose or Draw	Syndicated (Buena Vista)

- **People:** non-celebrity individuals or couples reveal embarrassing personal details about themselves, as in "The Dating Game" and "Love Connection."
- **Stunts:** people do odd things for money and prizes, as in"Beat the Clock," "Truth or Consequences," and "Let's Make a Deal."
- **Gambling:** contestants answer some preliminary questions so that they may earn the right to spin the wheel, roll the dice, reveal a card, and so forth. Examples of this program type include "Card Sharks," "High Rollers," and "Wheel of Fortune."
- **Charades:** celebrities and others make up teams that convey movie titles,

themes, books or phrases through pantomime, e.g., "Body Language," "Showoffs," or "Say It With Acting."[20]

Notes

1. Jefferson Graham, *Come On Down! The TV Game Show Book* (New York: Abbeville Press, 1988), p. 7.
2. *Ibid.*
3. Castleman and Podrazik, p. 15.
4. *Ibid.*, p. 34.
5. Graham, p. 18.
6. *Ibid.*, p. 23.
7. By this time, the DuMont network was no longer in existence.
8. Graham, p. 33.
9. "Jeopardy!'s" hook was the novelty of providing the answers and having contestants pose the questions. "Password" used celebrities to give one-word clues to one-word answers. "Video Village" was a life-sized board game, with contestants advancing on rolls of the dice.
10. On the rolling credits at the end of each show, the producers acknowledge that some of the celebrities have been given some answers before the show to add to the overall quality of their responses.
11. Original debut, 1957.
12. Original debut, 1961.
13. Original debut, 1953.
14. These changes included such things as new hosts, new game boards, new graphics, and perhaps an added assistant.
15. Graham, p. 42.
16. *Ibid.*, p. 43.
17. National Association of Television Program Executives.
18. John Dempsey, "Syndies See '90 as Year of the Gameshow," *Variety* (September 13, 1989), p. 59.
19. *Ibid.* p. 61.
20. Graham, p. 54.

4

▼ Soap Operas

Soap operas have been a staple in American broadcasting since the early 1930s, and are likely to continue as such for a very long time to come. Today, an estimated 50 million people are considered to be regular soap opera viewers,[1] including two-thirds of all American women in households with television. This dramatic figure translates into more than $900 million in total network revenues each year—or one-sixth of all annual network profits.[2]

Initially, the notion of serial drama was considered risky and programmatically unsound. Most network executives felt that listening audiences would accept only those storylines that were resolved within a series episode. However, in the early 1930s, programmers experimented with an open-ended evening comedy called "Amos 'n' Andy." This show, along with serial comedies like "The Goldbergs " and "Myrt and Marge," became popular immediately, proving that the serial could be a successful form of radio entertainment.[3] However, despite the demonstrated popularity of open-ended storylines, network programmers were still hesitant to move prime-time serial form to daytime audiences. Their concerns centered around the seemingly unattractive listening population (housewives) during this time block, and the questionable cost efficiency of providing serious drama in continuous segments. Despite these reservations, however, the networks decided to experiment with several 15-minute episodes, provided at discount prices to interested sponsors in the early 1930s. Most advertising support for these daytime dramas came from corporations like Colgate Palmolive-Peet and Procter and Gamble, who sold household products to interested female listeners. Thus, the term *soap opera* was developed, to describe the melodramatic plotlines sold by detergent companies.[4]

The introduction of daytime drama met with as immediate a success as had its evening serial counterpart. Devoted listeners faithfully followed the lives and loves of their favorite soap opera characters. And, much to the networks' surprise, housewives were not an unattractive listening demographic to possess. In fact, programmers soon discovered that housewives, while not directly in the labor force, often controlled the purse strings of the household economy. Accordingly, by 1939, advertising revenue for the popular serials had exceeded $26 million.[5] Housewives had indeed found an alluring substitute for previous programming fare (such as hygienic information, recipe readings, and household tips), and were demonstrating their consumer power as well. Network programmers and advertisers had inadvertently stumbled onto an undiscovered gold mine. However, creative programming was not the only reason for the immediate popularity of radio soap operas. To better under-

Soap Operas—1991 Availabilities
First-Run Network Programs

Program	Distributor
All My Children	Network (ABC)
Another World	Network (NBC)
As the World Turns	Network (CBS)
The Bold and the Beautiful	Network (CBS)
Days of Our Lives	Network (NBC)
General Hospital	Network (ABC)
Generations	Network (NBC)
The Guiding Light	Network (CBS)
Loving	Network (ABC)
One Life to Live	Network (ABC)
Santa Barbara	Network (NBC)
Tribes	Network (Fox)
The Young and the Restless	Network (CBS)

Current Off-Network Reruns

Program	Distributor
Dallas	Syndicated (Warner Bros.)
Dynasty	Syndicated (20th Century Fox)
Falcon Crest	Syndicated (Warner Bros.)
Knots Landing	Syndicated (Warner Bros.)
L.A. Law	Syndicated (20th Century Fox)
St. Elsewhere	Syndicated (MTM)
Thirtysomething	Syndicated (MGM/UA)
Wiseguy	Syndicated (Stephen Cannell)

stand the success of daytime drama in the 1930s, it is important to look at two additional factors: the story formula and its relationship to post-Depression America.

THE PLOTLINES

The assembly-line approach towards developing the soap opera genre we know today began in the early 1930s, when many of the first serials were created by two advertising executives, Frank and Anne Hummert. They originated many of the popular early daytime dramas like "Just Plain Bill," "The Romance of Helen Trent," and "Ma Perkins."[6] The Hummerts based most of their stories in the Midwest—an ideal setting for several reasons. First of all, the Hummerts' ad agency was located in Chicago. They felt that their soap operas should be produced there because as the creators, they would be conveniently located, and could produce programs less expensively than in New York.[7] After making this decision, it followed that the Hummerts would choose much of their talent from the immediate area, thus creating more authentic drama from the Midwest than from other areas of the country. Finally, the Hummerts felt that the Midwest carried with it an accurate reflection of American values, attitudes, and lifestyles. It seemed to be an ideal part of the country for audiences to associate with the familiar themes of daytime drama, known as the *Hummert formula.*

The Hummerts' story formula was really quite simple: they combined fantasies of exotic romance, pathos, and suspense with a familiar environment of everyday life in a small-town or rural setting. Combined with an identifiable hero or heroine, this formula produced an overwhelming audience response. For example, each episode of "Our Gal Sunday" began with the following.

> (Voiced over the song, *"Red River Valley"*):
> *"Our Gal Sunday"*—the story of an orphan girl named Sunday from the little mining town of Silver Creek, Colorado, who in young womanhood married England's richest, most handsome lord, Lord Henry Brinthrope. The story asks the question: Can this girl from a mining town in the west find happiness as the wife of a wealthy and titled Englishman?[8]

Listeners flocked to hear the adventures of the small-town girl they felt they knew and loved in predicaments they might only experience vicariously. While the settings and characters varied from soap to soap, the underlying premise in Hummert daytime drama was always the same: that people everywhere shared common needs, common values, and common problems.

This broadcast unity of beliefs and attitudes was especially important during the post-Depression era, when poverty, unemployment, and general political pessimism threatened the very fiber of American family life. Listeners found courage through soap opera characters such as Ma Perkins and "Just Plain Bill" Davidson—common folks who could survive despite overwhelming odds. Their victories over the trials and tribulations of daily living gave many Americans the feeling that they, too, could and would survive.

Another formidable contributor to early soap opera formula and content was Irna Phillips, whose style differed from the Hummerts' somewhat, but was equally successful nonetheless. Phillips rose to fame as the creator of radio's first serial, "Painted Dreams," and was an equally powerful force in the development of many serials thereafter. She concentrated on characterization more than on plotline fantasy (as in the Hummert dramas); and later became noted for introducing "working professionals," that is, doctors and lawyers, to daytime serials. Phillips' approach to characterization was not always as popular as the Hummert storyline fantasies in the 1930s. However, in the early 1940s, when post-Depression escapism was not so necessary in America, her more realistic approach to daytime drama caught on quickly. In addition to "Painted Dreams "(1930), Phillips' credits during the early days of soap opera history included "The Guiding Light" and "The Road of Life" (1937), "Woman of White" (1938), and "The Right to Happiness" (1939).

It can be said that daytime drama entered the decade of the 1940s with several firmly established tenets of serial writing. First, characterization was simple, straightforward, and easily recognizable.[9] Second, characters found themselves in predicaments that were easily identifiable by their listeners, with settings easily imaginable to those who had never travelled far beyond their home environment. As Rudolf Arnheim discovered in his study, "The World of the Daytime Serial,"[10] soap opera characters seemingly preferred everyday occurrences in their own home town, as opposed to problems in an unknown environment. When circumstances necessitated travel, the new setting invariably took place in the United States. Arnheim surmised that soap opera producers refrained from international travel because they felt that listeners would not enjoy a foreign setting, which would demand that they imagine a place outside of their own realm of experience.[11]

Third, most of the action centered on strong, stable female characters, who were not necessarily professionals, but community cornerstones, nonetheless. Men were very definitely the weaker sex in soap opera life—a direct reflection on the primary listening audience during the daytime hours.

Finally, daytime drama was often used as a vehicle for moral discussions or a rededication to American beliefs and values. Soap opera heroines often voiced the platitudes of the golden rule as well as the rewards that would come to those who could endure the trials and tribulations of living in a troubled society. Take, for example, Ma Perkins' philosophy in 1938:

> Anyone of this earth who's done wrong, and then goes so far as to right that
> wrong, I can tell you that they're well on their way to erasing the harm they did
> in the eyes of anyone decent.[12]

While it is true that the basic soap opera formula was established during the 1930s, other important characteristics present in today's soaps evolved during the 1940s, 1950s, 1960s, and 1970s. For example, toward the end of the 1940s, crime emerged as an important plotline theme, especially in the area of juvenile delinquency. This direction was reflective of the times, for Americans were becoming increasingly concerned about youth crime in their country. Criminal story lines continued throughout the early 1950s and are still an important theme in today's daytime drama (although the situations have been updated considerably).

In the early 1950s, most soap operas moved from radio to television, and the resulting change in technology was felt at all levels, including scriptwriting. More specifically, daily serials soon became 30 minutes in length as compared to the 15-minute capsules of the 1930s and 1940s. Because viewers could now see their characters, plotlines became more slowly paced to capitalize on the advantages of the visual medium, such as character reactions and new locales. In fact, a common plotline, like a marriage proposal, could last for weeks in a 1950s' television soap. After the male character popped the question, several days of programming would be spent learning the reactions of both the principal and supporting characters to this event: the bride-to-be, her mother, her old boyfriend, his old girlfriend or ex-wife, his secret admirer, her secret admirer, etc., etc. The possibilities were endless. Thus, one major plotline could sustain itself for weeks longer on television than would have been possible on radio despite the added 15 minutes of programming each day.

Relatedly, the visual medium of television allowed for a wider choice in soap opera settings, because writers were not forced to limit themselves to the experiential world of radio listeners. Instead, they could take their characters anywhere, as long as they visually established the appropriate setting. However, the visual element in television also had distinct limits: soap writers could no longer *rely* on imagination to set a scene. Rather, they had to create the mood visually—a time-consuming and potentially expensive venture. As a result, stories with more true-to-life settings, such as "The Guiding Light" and "Search for Tomorrow" became more popular to both networks and viewers.

The late 1960s has often been described as the "era of relevance," due mainly to Agnes Nixon's contributions to soap opera drama on the ABC network.[13] Nixon became a forerunner in including different racial and ethnic groups in her soaps; and these characters were often confronted with such societal problems as drug addiction, child abuse, and venereal disease as well as the more traditional fare of love, marriage, and children. Sex was freer, women were more independent, and community problems were addressed more directly than in the past. This new programming trend was lauded by both viewers and critics: ratings skyrocketed and scholars now asserted that daytime drama was both entertaining and informative. In fact, much of the research in the early 1970s centered on the positive educational effects of watching soaps experienced by viewers.

In the late 1970s and the 1980s, other factors, such as prime-time television, the film industry, and television rating structures, have dictated the way daytime serial plotlines are written. Over the last several years, for example, storyline segments in most daytime serials have imitated scenes from such prime-time television fare as "Dynasty" and "Knots Landing," and movies such as "Flashdance," "Fatal Attraction," "Raiders of the Lost Ark," and "Witness." Since prime-time series and movies are much more expensive to produce than soap operas, it's obvious that there have been limitations to daytime serial imitation. However, soap opera producers seem to feel that success in any context is worthy of replication; and this feeling has been most directly conveyed in daytime drama through longer program segments (changing from half-hour to hour-long shows) and larger budgets over the last several years.

Daytime drama has also adopted a prime-time attitude toward ratings and their importance to program survival. Thus, it is not very surprising to learn that soap operas save most of their key dramatic action for ratings sweep periods, much like their counterparts in prime-time drama. In fact, the intelligent viewer's key to watching soap operas over the past several years has been to concentrate on programs during the months of February, May, July, and November—the sweep periods—because, as in prime-time shows, more action, adventure, passion, and problem resolution are likely to occur during these times.

In reviewing the types of plotline themes between 1983–1990,[14] it is important to note several factors as they relate to television ratings and the sweep periods. First of all, the story lines most likely to be aired during a ratings sweep period include the following:

1. Marriage
2. Investigations (usually involving international travel)
3. The discovery of key clues in some type of crime or mystery
4. Court cases
5. Criminal arrests

6. Dramatic accidents such as fires, explosions, or auto crashes
7. The discovery of a character with a false identity
8. Life-saving surgery
9. Death

Also, it is not unusual for a person suffering from a traumatic injury or psychosomatic illness to recover dramatically in February, May, July, or November.

In contrast to sweep period story lines, certain themes occur consistently throughout the year, thus classifying them as part of the general soap opera formula. Not surprisingly, these themes have been characteristic of daytime drama since its inception in the early 1930s. These themes are the following:

1. Parenthood
2. Romance
3. Love
4. Jealous lovers
5. Romantic stumbling blocks
6. Romantic break-ups
7. Passion
8. Psychological problems
9. Crime
10. Deception
11. Jobs
12. Money

Clearly, it can be said that soap opera plots in the 1990s still revolve around issues of love, family, health, and security—much like the storylines in early radio daytime drama. However, ratings sweep periods now play a significant role in story selection as well as character creation and elimination. In short, an actor can project her or his future to last no longer than the next ratings sweep. These periods represent a time of conclusion as well as a time of renewal.

THE AUDIENCE

There is no question that the common thread linking most soap opera listeners in the early 1940s was gender. While differing in age, economic class, lifestyle, and geographic location, the fact remained that women were the greatest fans of radio serial drama. However, as researchers investigated the audience demographics more carefully, they found that a more describable soap opera follower did exist—one definable in terms more specific that "female."[15]

One clearly notable observation about most daytime listeners in the early 1940s was that they had been measurably affected by the Depression in the late 1920s and early 1930s. The effect was usually indirect, however, since women of that time were not a large factor in the workplace. Instead, their husbands, fathers, sons, or brothers were often unemployed; and the women were left to compensate for their income loss at home. The few women who found work outside the home were usually single, and not career-oriented. Their primary goal was to get married; they generally worked at their jobs until they found husbands and had children. Thus, it is safe to say that the typical soap opera listener was married and part of a single-income household. Author Robert Allen notes the following:

> The job of the typical woman in the 1930s was enormous but clearly defined: it was her task to "keep things going," to hold family and home together against the economic ravages of the Depression, to minimize the deterioration in living standards most families suffered. For all the talk of flappers and changing roles for women during the 1920s, what carried over into the 1930s was the division of family, social, and economic roles according to sex.[16]

Further, according to Herta Herzog,[17] the typical serial listener of the 1940s did not have a formal education beyond high school; in fact, many of the listeners surveyed had not gone past elementary school. Non-listeners, on the other hand, seemed to graduate from high school and beyond more often. Despite the comparative gap in education between soap listeners and non-listeners, incomes between the two groups were not radically different. Both seemed to be affected by the post-Depression environment and were members of single-income households.

Herzog also found that most listeners lived in rural settings rather than in large metropolitan areas or small cities. She reasoned that one possible explanation for this phenomenon was that more citified environments provided more activity choices for people. Thus, they were able to choose other things over listening to the radio.

Finally, Herzog developed a *sophistication index* to compare serial listeners with non-listeners. This sophistication index was made up of various components, such as an interest in reading, owning a telephone,[18] and participation in outside activities and hobbies. After tabulating the results, Herzog found that reading between the two groups did not vary greatly. However, the types of preferred books and magazines differed between listeners and nonlisteners. Listeners read more mystery novels, while nonlisteners enjoyed historical novels. Also, Herzog observed that serial listeners tended to spend more time at home, preferring vicarious experiences to those that were firsthand.[19]

Thus, in summary, the typical radio soap opera listener of the early 1940s was a married female, between the ages of 18 and 35, living in a rural area, with a high school education or less. She also enjoyed activities at home more than outside entertainment or other hobbies.

A similar study conducted in 1985 reflects a much more varied audience composition.[20] While it is still true that most soap opera fans are female, further descriptions of economic class, education, and sophistication are more difficult to make than in years past (see Tables 3 and 4). For example, it's very clear that educational levels and economic situations have changed greatly since the early 1940s. In addition, while today's typical soap fan is still female and usually between 18 and 35, many daytime serial viewers are younger, male, more career-oriented, more highly educated and, in general, more diverse than their earlier counterparts. Thus, soap operas are now intended to attract specific target groups.

With this tendency in mind, the 1985 study carefully defined specific audiences for most of today's soap operas.[21] It is important to note that the following comments serve only as partial audience descriptions. This is because many people have very

▶ **Table 3** Age and economic class breakdowns for the 1985 study

Gender/Age	Lower	Lower-Middle	Middle	Upper-middle	Upper
Males 1-10 (12)	8.3%	8.3%	50%	25%	8.3%
Females 1-10 (12)	8.3%	8.3%	66.7%	8.3%	8.3%
Males 11-20 (59)	5.1%	3.4%	47.5%	39%	5.1%
Females 11-20 (128)	1.3%	10.9%	46.2%	40.3%	1.3%
Males 21-40 (93)	5.4%	8.6%	44.1%	38.7%	3.2%
Females 21-40 (156)	1.3%	10.9%	46.2%	40.3%	1.3%
Males 41-60 (36)	-	11.1%	41.7%	41.7%	5.5%
Females 41-60 (63)	1.5%	4.8%	50.8%	33.3%	9.6%
Males 61+ (16)	6.3%	12.5%	50%	3.2%	-
Females 61+ (25)	3.8%	11.5%	46.2%	38.5%	-

NOTE: The numbers in parentheses indicate the actual numbers of people surveyed in each group.

personal reasons for identifying with certain soap operas. They are not limited to demographic statistics. It also should be noted that the viewer comments were gathered during the 1983-1985 seasons; since then, network programmers have made major and minor changes in soap opera storylines and characterizations. Changes are made because network executives constantly seek to maintain a certain set of audience demographics or try to shift from one age/gender group to another. As indicated earlier in this chapter, ratings are the name of the network-advertiser game, so the quest for preferable demographics is everpresent. Relatedly, preferable demographics are no longer defined only as the female age group between 18 and 35. For example, advertisers have noted recently that men and women over 50 usually possess a relatively large disposable income as well as more leisure time in which to spend it. Accordingly, this age group has become a very attractive demographic to hold. Thus, the strategy of appealing to two totally different age/gender groups (as in "Days of Our Lives") may become a more popular trend in the future.

In any case, the following comments reflected audience preferences in eight of today's most enduring soap operas:

"All My Children"

This soap was extremely popular with both men and women in the college-aged and the 21 to 40 age groups. Females aged 41 to 60 showed some interest in

▶ **Table 4** Age and educational breakdowns for the 1985 study

Gender/Age	Elementary School	High School	Community College	College	M.A.	PH.D
Males 1-10 (12)	100%	-	-	-	-	-
Females 1-10 (12)	100%	-	-	-	-	-
Males 11-20 (59)	3.4%	33.9%	3.5%	59.3%	-	-
Females 11-20 (128)	9.4%	18%	4.7%	67.9%	-	-
Males 21-40 (93)	-	4.3%	6.5%	75.3%	11.8%	2.1%
Females 21-40 (156)	-	10.3%	2.6%	73.1%	12.8%	1.2%
Males 41-60 (36)	5.6%	33.3%	-	38.9%	13.9%	8.3%
Females 41-60 (63)	-	28.6%	12.6%	49.2%	4.8%	4.8%
Males 61+ (16)	12.5%	62.5%	12.5%	12.5%	-	-
Females 61+	11.5%	42.4%	19.2%	19.2%	7.7%	-

NOTE: The numbers in parentheses indicate the actual numbers of people surveyed in each group.

it, but the overwhelming majority were men and women aged 18 to 35. When asked why they enjoyed it, common responses included the fact that "All My Children" combines serious themes with a sense of humor. Viewers also liked the various ethnic and economic groups as well as the mix between traditional and contemporary issues.

"Another World"

Statistically, "Another World" seemed most popular with males and females in the older demographic groups (50 and over). However, due to the fact that only a small number of people surveyed claimed to watch this soap opera, this analysis may be misleading. In any case, "Another World" was characterized as very traditional in nature. Most popular themes at the time included romance, drugs, criminal investigations, job-related problems, and illness/injury.

"As the World Turns"

Not many viewers surveyed listed this soap as one of their favorites, but of those who did, the over-50 age group was by far the leading viewership. Of those interviewed, many people had been fans of "As the World Turns" since its first year; and felt it was a very strong daytime drama. They also felt that the soap was very traditional in nature.

"Days of Our Lives"

The two age groups most devoted to this soap were male college students and females aged 41 to 60. It seemed that "Days of Our Lives" appealed to two different audiences: one that was made up of young males, interested in themes such as investigations of exotic crimes and action and adventure; and another audience of older women who enjoyed romance, love, and marriage. Also, it was interesting to note that many of the younger viewers watched "Days of Our Lives" and "General Hospital" each day because of similar story lines.

"General Hospital"

After 22 years, "General Hospital" was still a very popular soap opera in 1985. High percentages in all age groups from 11 to 60 listed this drama as one of their favorites; in fact, female college students and males between ages 21 and 40 listed it most often as their top choice. However, some of the people interviewed at this time indicated that they were becoming disenchanted with "General Hospital" because of increasingly unbelievable storylines and slow plotline development. Instead, they preferred themes dealing with international travel, criminal investigations, and job concerns.

"The Guiding Light"

"The Guiding Light" seemed to collect audience support from all age groups, with females over 50 registering as the most numerous viewership group. The latter figure may have been due to the fact that this is the longest survivor in daytime drama. Having started in radio in 1937 (and moving to television in 1952), "The Guiding Light" is still going strong today. Some of the younger persons interviewed indicated that they had recently switched from "General Hospital," preferring "The Guiding Light's" pace and storylines. Other, more mature people said that they had "grown up" with the Bauer family and had never left "home."

"One Life to Live"

While the percentages were somewhat lower than in "All My Children" and "General Hospital," "One Life to Live" held considerable popularity with men between ages 18 to 35. Several persons indicated that they had really started watching this soap opera because it aired between "All My Children" and "General Hospital." However, they now listed it as their favorite daytime drama.

"The Young and the Restless"

"The Young and the Restless" seemed strongest in viewer appeal for women in the 21 to 45 age group. When asked about this soap, most said they like the wealth, glamour, and fantasy-like lives of the characters. (This, by the way, was in direct opposition to those who liked "All My Children.") Popular plotline trends included pregnancy and parenthood issues, romantic rebounding, heroes versus villains, and high-powered careers and finance.

The overwhelming popularity of today's soap operas suggest that they still fulfill certain needs for the viewing public. According to the 1985 audience survey, these needs include emotional release, fantasy, advice, and certain types of information. The following points also serve to detail viewer comments from the survey:

1. Overall, viewers claimed to choose specific soap operas because they enjoyed seeing certain characters' reactions to the situations in which they found themselves. Plotlines were the second most important reason for watching daytime drama.

2. Heavy soap opera viewers (those who regularly watched three or more programs) seemed to be loyal to specific networks; that is, they tended to watch all ABC, all CBS, or all NBC dramas. However, many of those who watched only two soaps chose them because of specific story line content, pacing, or characterization, despite differences in networks. One of the best examples of this phenomenon was the popular dual combination of "Days of Our Lives" (NBC) and "General Hospital"(ABC). Another popular mix was "All My Children"(ABC) and "The Guiding Light" (CBS).

3. Timeslots for competing soap operas were no longer considered to be as important an issue as in the past, largely because of the increasing amount of VCR use. Many of those interviewed said that they regularly viewed two soaps aired at the same time, either by watching one while taping the other or by having friends tape them. No longer are soap operas available only to those who are home during the day.

4. Audiences were very loyal to their specific soap operas, many having watched the same ones for years and years. However, loyalty to past habits was not the main reason for their continued viewing. As a matter of fact, some persons stated that while they grew up with one soap, they switched to another because they preferred the latter drama's characters and plots. In the last analysis, audiences were most dedicated to programs that provided good storylines and characters; they did not remain loyal for long if they were disappointed continually.

5. While the 1985 survey showed that ABC led in overall numbers, the other two networks claimed equally devoted fans. And soap operas were very definitely an intended choice for viewing (whether on ABC, CBS, or NBC). With so many programming alternatives available through cable, satellite, and home video, audiences were not forced to select daytime drama because it was the least objectionable choice. They were happy to be active, dedicated fans of this genre.

In summary, soap opera fans have grown more sophisticated, in keeping with the nation's economy, educational system, and technology. They can no longer be quantified as a homogeneous group of married, middle-class housewives, with little or no formal education. However, despite their new sophistication, fans are still extremely zealous when discussing their favorite stories and the characters within them. For most viewers of daytime drama, fictional families like the Tylers, Quartermaines, Buchanans, and Bauers contain real friends, as well as foes.

At present, daytime television provides networks with 60 to 80 % of their annual revenues. Since much of the programming during this time block is composed of soap operas, it's fair to say that daytime drama supports many of the networks' prime-time shows.[22] For example, the average cost of producing one episode of "Dallas" is $1 million; an *entire week's* cost of a soap opera is between $300,000 and $400,000.[23] Unfortunately, however, most of the networks' highest moneymakers often face severe limitations in program quality. In the case of daytime soaps, these limitations are usually reflected in budgetary and content restrictions.

Economically speaking, the allowance for most soap opera productions is based on a low investment/high yield principle. As a result, salaries for most writers and actors on daytime television are considerably lower than for their counterparts in prime time. Also, the numbers of characters and on-location scenes in each show are much more limited in afternoon soaps as compared to most nighttime series.

Since the monetary investment is so high in prime-time television, networks cannot afford to support a prime-time program while it goes through a dry spell of mediocrity. Consequently, shows such as "Falcon Crest" and "The Colbys," are more apt to face extinction than a soap opera like "Loving" or "Another World," even though each of the latter has had more significant declines in ratings over the years. In short, daytime serials may have smaller budgets with which to work, but they also enjoy greater network patience over longer periods of time than nighttime soaps.

Soap operas are unquestionably one of the most recognized genres on TV today. Like their radio counterparts, they enjoy consistent audience devotion and popularity in a world where most success is as ephemeral as the last ratings period. Indeed, America's romance with daytime drama has endured for a very long time, and based on all the existing research available, the future of the afternoon soap is as secure as its past.

Notes

1. *Regular soap opera viewers* are defined as those persons who watch at least one soap at least three times each week.
2. Robert C. Allen, *Speaking of Soap Operas* (Chapel Hill, NC: University of North Carolina Press, 1985), p. 3.
3. Robert LaGuardia, *Soap World* (New York: Arbor House, 1983), p. 9.
4. Muriel Cantor and Suzanne Pingree, *The Soap Opera* (Beverly Hills, CA: Sage, 1983), pp. 36–37.
5. J. Fred MacDonald, *Don't Touch that Dial: Radio Programming in American Life from 1920 to 1960* (Chicago: Nelson-Hall, 1979), p. 233.
6. LaGuardia, p. 12.
7. *Ibid.*, p. 20.
8. MacDonald, p. 235.
9. *Ibid.*, p. 243.
10. Rudolf Arnheim, "The World of the Daytime Serial," in *Radio Research:1942–1943*, eds. Paul F. Lazarsfeld and Frank N. Stanton (New York: Essential, 1944), pp. 38–85.
11. *Ibid.*, pp. 38–39.
12. MacDonald, pp. 243–244.
13. La Guardia, p. 39.
14. Much of the basic research for this analysis comes from the author's earlier book entitled, *The Soap Opera Evolution: America's Enduring Romance with Daytime Drama* (Jefferson, NC: McFarland & Co., Inc., 1988). The data has since been updated and applied to the 1990 network soaps.
15. Allen, pp. 135-136.
16. *Ibid.*, p. 136.

17. Herta Herzog, "What We Know About Daytime Serial Listeners," in *Radio Research: 1942–1943*, pp. 3–33.

18. Herzog made an interesting observation while testing for phone use in her sophistication index. When comparing women of the same economic level who owned phones to those who did not, she found more serial listening was done by non-phone households. Because the Hooper ratings service conducted all of its interviews by phone, Herzog felt that the daytime serial audience was vastly underrated. (See Herzog, p. 9, footnote 3).

19. *Ibid.*, pp. 8–12.

20. To better understand today's soap opera audience, the author tried to replicate Herzog's methodology by conducting a random survey of 600 subjects during the last week of November, 1985. Interviewees were asked how much television they watched; which daytime programs they preferred; and, if they were soap opera viewers, why they chose to watch specific soap operas. In all studies dealing with randomly selected subjects, researchers run the risk of creating samples which are not totally representative. In this study, 600 people were randomly selected in several areas of the country. However, because college students served as the data gatherers, it is possible that the resulting sample was somewhat skewed educationally and economically. This does not take away from the fact that today's viewers are educationally and economically better off than in the 1940s—other studies have confirmed this trend. It is important to note, though, that the percentages may be higher in this analysis than in other research.

21. Several soap operas seen today were not included for various reasons. "Loving," according to the survey, showed no real distinction in any category. In fact, most people interviewed said they watched it because they were on lunch break or because they were waiting for "All My Children," "Santa Barbara" still seemed to be searching for its identity in 1985, having just begun in the summer of 1983. (Subsequent research, however, has shown that this soap attracts demographic groups similar to "Days of Our Lives") "The Bold and the Beautiful," and "Tribes" have debuted since the survey was taken.

22. Diane Haithman, "Soap Writers Rule Daytime TV with a Godlike Hand," *Detroit Free Press* (November 25, 1984), p. 1G.

23. These figures were gathered from a 1989-1990 list of production costs found in *Variety*.

5

▼ Children's Programming
▼
▼
▼
▼

As evidenced by early broadcasting schedules, TV programmers have always perceived children to be a significant and well-defined viewer demographic. The importance of children to advertisers was much less apparent at first, however. Potential sponsors saw this age group as inattentive, fickle, and economically powerless. In fact, only after several years of proven ratings success did retailers and manufacturers (as well as their advertising agents) recognize the marketing potential of younger audiences. They later applied this research to character promotion, product placement, and of course, thousands and thousands of commercials.

Serious interest in children's programming took hold less than a year after TV's formal introduction to the American public (in 1948). Before this time most of the available broadcast hours were devoted to sporting events such as boxing, wrestling, football, and baseball. The primary rationale behind this decision was that the greatest number of receivers in the United States were located in neighborhood pubs and bars at this time. In 1949, however, the number of television sets manufactured had increased dramatically, and over 100,000 Americans enjoyed TV in their own households. Accordingly, during that same year, over one-third of all broadcast time was given over to children's programming.

By 1951, the four major networks (ABC, NBC, CBS, and DuMont) were airing at least 27 hours of children's shows weekly, at times when the majority of that target viewing audience was available—weekday evenings between 6:00 and 8:00 P.M. TV quickly became the greatest source of entertainment for the young. According to a study conducted by Eleanor Maccoby in 1950,[1] for example, Boston Children (on average) watched about 2½ hours of television each weekday and 3½ hours on weekends. They stopped going to movies and were less likely to do their homework willingly. The children were also reportedly going to bed at least 25 minutes later than before the days of television. In addition, social distinctions had begun to emerge among those children with TV and those without: those children without TV waited impatiently for sets of their own, but they still managed to watch at other houses (most likely, in part, to keep up with all the current small talk among their peers).

At the time, Maccoby and others thought that the television frenzy would only be temporary, due to the novelty of the new medium. However, as later studies would indicate, the figures for TV viewing, in fact, increased.[2]

While television in the early 1950s gave a certain priority to kids shows like "Howdy Doody" and "Kukla, Fran and Ollie," these programs were often listed as *sustaining*, that is, presented without advertising support. The lack of sponsorship at this time was linked to the assumption that children, alone, were not powerful consumers. At best, they were seen only as a larger viewing audience. Youngsters were also considered potentially valuable as a persuasive means of adding a new television set to the household—but beyond that, they were decidedly innocuous.

Needless to say, however, this attitude did not last. Before long, market researchers saw children's television as having great potential for character and product promotion.

CHARACTER AND PRODUCT PROMOTION

Character promotion had been in existence long before television itself had become a reality. The first evidence of such a concept occurred in 1904, when the Brown Shoe Company purchased the rights to use the name of Buster Brown. Brown was a popular comic strip character of the time; and the Brown Shoe Company thought it might be advantageous to use him to promote its children's line of shoes at the Saint Louis World's Fair.

The idea of associating a character with a product became an instant hit! The trend spread rapidly, with such Hollywood luminaries as Charlie Chaplin and Shirley Temple franchising themselves in dolls, toys, books, and comic strips.[3] Marketing analysts first thought that the concept of character promotion would do little more than capture a bit of free advertising. They had no idea how far the strategy could go.

By the time character promotion had come to television, show characters were actually endorsing products during commercial breaks within the program. One of the earliest examples of this sales gimmick was "Howdy Doody's" Buffalo Bob, who urged his young viewers to "have your Mom or Daddy take you to the store where you get Poll Parrot shoes, and ask for your "Howdy Doody cutout!"[4] Another instance involved "Hopalong Cassidy" consumer items (like pajamas, wallpaper, candy bars, and watches), which earned profits of over $70 million annually in the 1950s.

Because of such blatant abuse of character promotion on children's television, certain activist groups emerged in the late 1960s and early 1970s to combat the networks and advertisers. Among these organizations, a Massachusetts-based volunteer group called Action for Children's Television (ACT) was most effective and impressive.

In 1970, ACT presented a set of proposals to Congress and to the FCC, asserting its mission to improve children's programming. ACT members felt that this goal could be accomplished by introducing several simple (albeit controversial) safeguards. They included:

1. The elimination of advertising in all children's programming (although underwriting would be permitted);

2. A directive to all kid show performers, banning them from mentioning any products, services, or stores in their programs; and

3. An FCC requirement for stations to set aside a minimum of 14 hours weekly for children's programming.[5]

The latter directive was even more specific. ACT suggested certain hours for certain age groups. The desired schedule was as follows: 7:00 A.M.–6:00 A.M. (preschoolers); 4:00 P.M.–8:00 P.M. daily, 9:00 A.M..–8:00 P.M. weekends (ages 6 to 9); and 5:00 P.M..–9:00 P.M. daily, 9:00 A.M.–9:00 P.M. weekends (ages 10 to 12).

The ACT proposal met with considerable furor from both broadcasters and advertisers. Both parties argued that commercials could also be seen positively; in fact, research had shown that product promotion on children's shows often created enlightened consumer behavior.[6] Further, they questioned the definition of *children's viewing hours*, claiming that no one could *assure* any times as potentially free from young audiences. Thus, advertising prohibitions should be conducted across-the-board, or not at all.

Ultimately, Congress, the FCC, broadcasters, advertisers, and ACT reached closure on some compromise measures, encompassing several general guidelines:

1. Programs like "Captain Kangaroo" no longer allowed their hosts to deliver commercials or product endorsements within the context of their show. Late-fringe or prime-time programs (like "The Andy Griffith Show"), however, were not required to observe this restriction, since the *primary* audience was not children.

2. Advertisers agreed to use more discretion when determining the content and placement of their commercials. For example, children's vitamin spots were no longer used on shows specifically geared toward kids because there was some concern over the potential to confuse a medicine with a candy. Also, visual disclaimers (such as a size distinction or a notice that batteries would not be included in the purchase) were posted more prominently than before.

3. Broadcasters promised to make a serious attempt to increase educational television in their schedules. *Educational*, however, was defined as a learning experience imbedded within the more overt entertainment function of the show. In the early 1970s, such programming included "Curiosity Shop," "Kid Power," "Make a Wish," "ABC After School Special," "Multipication in" (ABC); "Fat Albert and the Cosby Kids", "The CBS Children's Film Festival", "In the News"(CBS); and "Talking With a Giant" and "Watch Your Child" (NBC).

Several years after ACT's initial confrontation with broadcasters and advertising executives, it succeeded in limiting the commercial time accorded to children's programming in the Saturday morning daypart. However, for the most part, ACT (as well as other special interest groups) was considered to be a much less powerful force in the 1980s than it had been the decade before.

One of the most dramatic examples of decreased consumer group influence has been evidenced by the growingly corrupt use of character promotion in the 80s and 90s. Character licensing abuse actually began in 1980, when Kenner Toys created a television show for the sole purpose of advertising their product—"berry" nice "Strawberry Shortcake"[7]. Other companies, including Coleco, Mattel, and Hasbro, have since followed Kenner's lead by producing such programs as "The Cabbage Patch Kids," "Masters of the Universe," "Smurfs," and "Teenage Mutant Ninja Turtles."

Some consumer activists like Peggy Charren (the founder of ACT) point to the Reagan presidency and its attitudes on broadcast deregulation for the current ills of television:

> The Reagan administration is wholly to blame for everything rotten that's happened to children's television. They've literally wrecked it.[8]

Other, more moderate critics, such as Joan Ganz Cooney (the creator of "Sesame Street") felt that the responsibility does not rest with congressional representatives or media executives alone. She suggested that parents, grandparents, or other influential adults watch television along with their children, and discuss with them the themes and ideas brought up in each program.

> That's the key right there—watching TV with the little ones. Grandparents as well as parents should be discriminating as to what the child watches—and, of course, I'd recommend "Sesame Street" and "Mister Roger's Neighborhood." But it's important to share in viewing these shows, or any shows. If you can watch and discuss the programs with the child, he or she will learn much more than from watching alone.[9]

But whatever one's personal feelings may be regarding children's TV programming, the fact remains that the most vocal audience collective will generally determine the direction of the genre. Sometimes social action groups can influence media legislation. However, more often, viewers vote on a show's success via product purchases, celebrity recognition, and merchandising demands. As viewer preferences change, so also do the programs aired, as evidenced by an analysis of programming trends.

PROGRAMMING TRENDS

In the early and mid-50s, NBC became the network most interested in children's programming. Until 1957, it consistently put the greatest number of children's shows on the air. NBC also created the first children's program review committee at this time to ensure high standards of treatment for their shows. The committee was formed, in part, because of a 1954 Yale study done on TV violence. This study asserted that, while networks seemed to be producing impressive adult programs, children were not receiving appropriate attention in the same area.[10] In fact, kids shows were often seen as the most violent and offensive productions on the air. (Ironically, however, they were also some of the most highly rated programs on network schedules.)

In any case, after reviewing the Yale study (as well as reading and hearing other viewer complaints), all networks renewed their interest in family-oriented programs with such shows as "Lassie," "Superman," "Hopalong Cassidy," "Father Knows Best," "The Lone Ranger," "The Adventures of Rin Tin Tin," and "Make Room for Daddy," Most impressive among the prime-time fare at this time was "Disneyland" a co-production effort from ABC and the Walt Disney studios.

In the early 1950s, ABC was still woefully behind NBC and CBS in audience ratings. Its utter desperation over this situation finally forced the network to seek help from another source—Hollywood. More than its competitors, ABC needed help in program development and production to fill its nightly schedule. As a result, the network decided to do two things: in 1953 it merged with United Paramount Theaters for much-needed cash; and in 1954 it signed an agreement with Walt Disney to produce 20 weekly one-hour programs for the 1954–1955 season. In return, ABC would contribute heavily toward the construction of a Disneyland theme park in Anaheim, California.[11]

"Disneyland" premiered in October 1954, to critical rave reviews. From the very beginning, this show began to change the notion of the children's television marketplace by presenting information in a uniquely entertaining way:

> Drawing on more than twenty years of theatrical material, as well as new features shot specifically for the show, *Disneyland* presented a delightful combination of kiddie adventure yarns, travelogues, real-life nature stories, mildly educational documentaries, and classic Disney animation. It was the perfect family show and a resounding ratings success, providing ABC with its first top ten hit in five years.[12]

"Disneyland" won the 1954 Peabody award for educational excellence, and in 1955, it was presented an Emmy for best adventure series.[13] In addition, the unbeatable combination of TV show and theme park became associated as one entity. Thus, Disneyland became the most coveted wish on every child's travel list; and ABC enjoyed the popularity of both the series as well as specific character fads, such as Davy Crockett, Donald Duck, Mickey Mouse, and Pluto.[14]

During the morning and afternoon hours in the 1950s, kids were provided with quality material with shows like "Captain Kangaroo," "Ding Dong School," "Romper Room," and "The Mickey Mouse Club." This program selection had been much improved from earlier years, when old, outdated theatrical films and movie shorts filled the daytime airwaves. Critics applauded the networks' attempts to educate as well as entertain their young audiences.

In addition, CBS began experimenting with the Saturday morning time block as well as with made-for-TV animation. On December 10, 1955, "Mighty Mouse Playhouse" premiered to the delight of America's children. This bold move demonstrated that Saturday morning hours could be profitable for advertising revenues. It also inspired future cartoon programming in both daytime and prime time.

William Hanna and Joseph Barbera were among the first animators to compete with Walt Disney studios for cartoon programming. Their product was aesthetically less pleasing than the painstaking brush strokes from the Disney consortium; but it was also less expensive, and was almost as popular among TV's young demographic. Before long, "Tom and Jerry," "Gerald McBoing-Boing," "Yogi Bear," "Huckleberry Hound," and "Pixie and Dixie," adorned lunch boxes, bubble bath containers, and T-shirts all over America. Sponsors (and networks) quickly learned that cartoon characters could sell other products; thus, a plethora of animated programming filled television schedules on weekend mornings, weekday afternoons, and, in the 1960s, during the weeknight prime-time hours.

In 1960, Hanna and Barbera teamed up with ABC to air "The Flintstones," an animated version of "The Honeymooners." Clearly, the program was targeted toward adults, given its story premise and time slot (8:30 P.M.). As Castleman and Podrazik note, however, the program idea was problematic and ill-conceived for prime time:

The series held great promise that was never realized. All the appealing elements of *The Honeymooners'* characters were lost in transition to animation, and *The Flintstones* emerged as a dimwitted interpolation in a Stone Age setting. Fred Flintstone was noisy, boastful, and stupid. His neighbor, Barney Rubble, was a dolt. The interaction and scheming of the two lacked the wit, energy, humor, and deep affection of the Jackie Gleason-Art Carney original.[15]

Nevertheless, "The Flintstones" became a hit, and lasted a full six seasons on prime-time television before moving over to Saturday mornings.[16]

In addition to Fred, Barney, Wilma, and Betty, other adult cartoon characters emerged throughout the early 1960s (via Hanna-Barbera Productions). They were found on shows such as "Top Cat" (a take-off of "Sergeant Bilko"), "Calvin and the Colonel" (a variation of "Amos 'n' Andy"), and "The Jetsons" (a futuristic "Flintstones"/"Honeymooners"). Each of these shows made a successful transition to weekend mornings and weekday afternoons after their prime-time runs.

In addition to the adult cartoon fad, the 1960s were marked by a concentrated shift toward higher quality daytime programming for kids. Personality programs of the 1950s (like "Captain Kangaroo" and "Bozo's Big Top") continued to fill the airwaves with valuable ideas and consciousness-raising lessons. However, these shows were later overshadowed by a new concept in public television, "Sesame Street." After witnessing "Sesame Street's" immediate success, the commercial networks tried to counter it with productions such as "Fat Albert and the Cosby Kids" (to promote healthy social values) as well as "In the News" (to create an interest in current events).

Children's programming in the 1970s was reflective of an overall trend in network television to instill family values within a rapidly changing political climate. In keeping with such prime-time fare as "The Brady Brunch," "Little House on the Prairie," and "The Waltons," both network and local programmers attempted to design shows with social orientation as well as entertainment value. For example, network offerings like "Kid Power" and "Talking with a Giant" addressed value judgments and interpersonal relationships. As for local productions during this time, several examples are worthy of note.

- WRTV (Indianapolis) with "Uncle Uri's Treasure." This program focused on a world traveler who shared "gifts of knowledge" from wide-reaching areas of the globe. After airing these productions, however, "Uncle Uri's Treasure" continued in the youngster's learning process by providing follow-up projects in conjunction with the local library, museum, and high school.
- KPRC (Houston) with "Sundown's Treehouse." This show was created by children, for children. It aired such segments as field trips to Houston and interesting natural phenomena. Handicapped children were often featured as guests on the program.
- WCPO (Cincinnati) with "The Uncle Al Show." This program lasted over two decades, and was centered around five major characters and their relationships to each other. It also featured a live studio audience, comprised of at least 40 children.[17]

With such an obvious commitment to quality television, it was not surprising that adults were increasingly supportive of the medium. In fact, a 1975 Roper survey found that nearly two out of every three parents of children under 12 felt that preadolescent programming had improved. Unfortunately, however, the rise of show credibility in the 1970s was quickly dissipated by children's programming in the 1980s.

The early 1980s seemed ripe for serious concentration in character merchandising. In addition to the "Strawberry Shortcake" craze (mentioned earlier in this chapter), most toy manufacturers and all of the networks raided the airwaves with such product-oriented shows as "G.I. Joe," "The Cabbage Patch Kids," "He-Man and the Masters of the Universe," "She-Ra: Princess of Power," "The Transformers,"and "Care Bears." Needless to say, the programming was accompanied by as much trademarked paraphenalia as possible—all advertised during the commercial breaks of the shows.

For the most part, these animated series were generally low in story content and high in problem solution through product purchase. As Tom Engelhardt pointed out in his essay on "The Shortcake Strategy":

> A painfully thin story [was] simply interspersed with a series of embarrassingly direct demands or even orders about how you should act, feel, and be. Take cooperation, for instance. Everyone realizes that kids should cooperate, while any fool (and TV producers are no fools) with child development experts and extensive research departments to call on knows that if your show advocates cooperation you can't easily be faulted. And so if it happens that you'd prefer to sell groups of little bears/puppies/fairies/ponies, then how useful to have wholly admirable reasons for suggesting that a parent plunk down a couple of hundred dollars. What better than urging kids to get sharing and togetherness and cooperation by buying whole integrated, cooperative, loving sets of huggable, snuggable, nurturing dolls? ("Ten Care Bears are better than one," as one Care Bear special put it.)[19]

One of the reasons why this strategy was so successful was that the traditional structure of American family life had changed quite abruptly within a decade. With the emergence of more and more single parent households as well as the increase of two-income families, television also increased its role as a surrogate parent for many children. By the mid-80s, the average youngster watched almost 4 hours of TV each day. Child psychologist Dr. Ron Slabey noted the following:

> Whether or not a TV show is intended by the producer to be educational or not, it teaches the children anyway, by example.[20]

Thus, the new heroes, comforters, friends, and supporters for many children of the 1980s became available for certain prices, in limited quantities, at the neighborhood toy store.

Toy manufacturers reveled in their newfound gold mine. For several years, Christmas season shoppers were forced to decide among numerous "hot" items for their children, hoping all the while that the chosen toy would be available when the household budget allowed the purchase. G.I. Joes, Cabbage Patch dolls, Transformers, and He-Man action figures all contributed healthy profits to their manufacturers. However, in the late 1980s, the bubble started to burst, and toy sales dropped significantly.

As the market plummeted, so also did the advertisers' demands for animated children's programming.[21] In 1988, WLVI's general sales manager, Thom Neeson, described the decline in the following way:

> In 1984, our station's children's TV billings made up 32 percent of total
> business. That dropped to 27 percent in 1986, and I project it'll be at 20 percent
> in 1989—and going down.[22]

Television programmers, both network-affiliated and independent, sought answers to the seemingly inexplicable drop in cartoon interest. After conducting several independent research studies, they concluded the following:

- The overemphasis on merchandising may have contributed to the lack of interest in watching the actual cartoon. Most children interviewed felt that the story line, characterization, and animation quality were often secondary to the primary goal of selling merchandise.[23] (Incidentally, this reaction lends credence to the Harvard Business School study on child consumer behavior mentioned in footnote 6 of this chapter.)
- Children aged 2 to 8 years old seemed fairly homogeneous in their television demands, that is, they didn't require distinctly different shows to appeal to their specific age and gender. However, children aged 9 to 11 revealed special programming tastes. For example, girls from 9 to11 seemed to like soap operas, court shows, talk shows, game shows and sitcoms. Boys aged 9 to 11, on the other hand, liked unisex and dramatic animation; however, they also enjoyed more adult programming, especially court shows and sitcoms.[24]
- Over 85% of the children surveyed found that all adult program genres were "appropriate for their age or older"; less than 50 % found any of the animated programming to be appropriate to their age or older.[25]
- Kids of all ages were still using their television sets, but for activities other than watching broadcast programs; they viewed cable shows, rented videos, or played video games instead. In fact, some of the product merchandising mentioned earlier (such as Nintendo) actually negated TV viewing in the long run, because children were using their TV sets to play the game.

After analyzing the findings, many broadcast programmers decided to change some of their traditional strategies for the weekday late afternoon daypart. For example, several stations chose to trim the kids' time period from 4 hours each day to 3, using the final hour for more *kiddult*[26] fare, such as off-network sitcoms or first-run court shows.[27] Other stations elected to eliminate the *kidvid* time slot altogether, filling the airtime with first-run talk shows, game shows, or off-network reruns. Still others decided to combine animation with live-action programming, and chose quality cartoons like "Ducktales" and "Chip and Dale's Rescue Rangers," as well as youth-oriented game shows like "Fun House."

By 1989, broadcasters were carefully rebuilding their programming strategies and hoping that the kids' market would resurface. They became pleasantly surprised to learn two significant facts.

First of all, some of the national ratings service reports were later found to be grossly inaccurate. The subsequent misinformation was later tied to the flawed performance of the people meter.[28] According to Michael Mellon, vice president of research for Buena Vista Television, there was a 30% discrepancy between information provided by the diary method versus the people meter measurement for certain individual programs. More specifically, the diaries from Nielsen during the 1988 November ratings sweep period reported that "Ducktales" reflected a 1083 VPVH (viewing per thousand viewing households) for children aged 2 to 11; while in people meters over the same time period, the show's rating was a 761 VPVH for the same audience.[29]

Probably the best reason for this discrepancy has been provided by a study done by J. Walter Thompson (the ad agency). Thompson researchers surveyed children about their preferences in using various electronic aids. They discovered the following:

> . . . not only could children operate microwave ovens, remote-control devices and VCRs, they could also operate a people meter. But when they use a microwave, the result is something good to eat. When they operate a remote control or VCR, they get movies and TV programs. But there is no tangible reward for using a people meter. In interviews, children told researchers that using a people meter would be fun for a while, but that they would probably tire of it or forget to use it.[30]

This new perspective on the ratings numbers gave networks renewed confidence to continue their previous commitments to children's programming. The "Big Three," however, decided to modify their previous revenue approach. Instead of accepting *barter only* deals (like those in the early 1980s), they demanded *cash only* or *cash-barter* arrangements. This adjustment in money collection guaranteed the funding broadcasters needed for their network revenues.

The other significant information for kidvid programmers in 1989 was the unexpected financial success of "Who Framed Roger Rabbit?," 1988's runaway theatrical movie hit. "Rabbit's" box office receipts ($153 million in 1988 alone)[31] proved that cartoon-based entertainment could continue to be profitable. It also unearthed a potential audience of adult animation viewers, provided the product was high enough in quality. Thus, by the 1990–1991 season, cartoon programming, which had previously been restricted to Saturday mornings and weekday afternoons, began to ease into prime time, with shows like "The Simpsons" and "Family Dog." This programming approach had not been used since the introduction 30 years earlier of "The Flintstones" in 1960.

Several media theorists searched for reasons behind the new animation craze.

> . . . the pop-sociological view [was] that aging baby boomers [were] nostalgically seeking to recapture the cartoon memories of their childhood—while keeping their own kids entertained.[32]

Thus, children's programming for the 1990s is likely to contain more "kiddult" fare than has been shown in the past 20 years. This trend, however, like those before it, will be subject to change according to the evolving wants and needs of television's youthful audience.

Children's Shows—1991 Availabilities
Children's Animation[33]

Program	Distributor
Adventures of Raggedy Ann and Andy	Network(CBS)
Alf	Network(NBC)
Alf-Tales	Network(NBC)
Alvin and the Chipmunks	Syndicated(Warner Bros.)
Animated Classics	Syndicated(Taffner)
Beetlejuice	Network (ABC)
Bill and Ted's Excellent Adventure	Network (CBS)
Bugs Bunny, Porky Pig	Syndicated (Warner Bros.)
Bugs Bunny and Tweety	Network (ABC)
Bullwinkle	Syndicated (DFS)
C.O.P.S.	Syndicated (Claster)
California Raisins	Network (CBS)
Camp Candy	Network (NBC)
Captain N: The Game Master	Network (NBC)
Care Bears	Syndicated (SFM)
Casper the Friendly Ghost	Syndicated (Worldvision)
Chipmunks	Network (NBC)
Chip n' Dales Rescue Rangers	Syndicated(Buena Vista)
Dennis the Menace	Syndicated (DFS)
Denver the Last Dinosaur	Syndicated(World Events)
Disney Afternoon	Syndicated (Buena Vista)
(4 variations)	
Disney's Gummi Bears,	Network (ABC)
Winnie the Pooh	
Ducktales	Syndicated (Buena Vista)
Dudley Do-Right	Syndicated (DFS)
Felix the Cat	Syndicated (Columbia)
Flintstone Kids	Network (ABC)
Flintstones	Syndicated (DFS)
Funtastic World Hanna-Barbara	Syndicated (Worldvision)
G.I. Joe	Syndicated (Claster)
Gravedale High	Network (NBC)
Gumby (new)	Syndicated (Warner Bros.)
Gumby (original)	Syndicated (Ziv. Intl.)
Heathcliff	Syndicated (Lexington)
Hercules	Syndicated (Columbia)
Inch-High Private Eye	Syndicated (DFS)
James Bond Jr.	Syndicated (Claster)
Jem	Syndicated (Claster)
Jetsons	Syndicated (Worldvision)
Jim Hemson's Muppet Babies (new)	Network (CBS)
Karate Kid	Network (NBC)

Children's Shows—1991 Availabilities (continued)
Children's Animation

Program	Distributor
Kissyfur	Network (NBC)
M.A.S.K.	Syndicated (Lexington)
Maxie's World	Syndicated (Claster)
Merry Melodies	Syndicated (Warner Bros.)
Mighty Mouse and Friends	Syndicated (Viacom)
Muppet Babies	Syndicated (Claster)
My Little Pony and Friends	Syndicated (Claster)
New Adventures	Syndicated (Lexington)
New Archies	Syndicated (Claster)
New Three Stooges	Syndicated (Muller Media)
Police Academy	Syndicated (Lexington)
Pup Named Scooby Doo	Network (ABC)
Real Ghostbusters	Syndicated (Lexington)
Rocky and Friends	Syndicated (DFS)
Scooby Doo	Syndicated (DFS)
Slimer! and the Real Ghost Busters	Network (ABC)
Smurfs (new)	Network (NBC)
Smurfs	Syndicated (Worldvision)
Snorks	Syndicated (Worldvision)
Space Kidettes	Syndicated (DFS)
Super Mario Brothers	Syndicated (Viacom)
Super Sunday	Syndicated (Claster)
Superfriends	Syndicated (Lexington)
Teenage Mutant Ninja Turtles	Syndicated (Group W)
Tennesse Tuxedo	Syndicated (DFS)
Thunderbirds	Syndicated (ITC)
Tiny Toons Adventures	Syndicated (Warner Bros.)
Tom and Jerry	Syndicated (Turner Program Sales)
Tom and Jerry Kids	Syndicated (Turner Program Sales)
Uncle Waldo	Syndicated (DFS)
Valley of the Dinosaurs	Syndicated (DFS)
Visionaries	Syndicated (Claster)
Wake, Rattle and Roll	Syndicated (Worldvision)
Wheelie and the Chopper Bunch	Syndicated (DFS)
Widget	Syndicated (Calico)
Yogi Bear	Syndicated (Worldvision)
Young Samson	Syndicated (DFS)

Children's Live Action

Program	Distributor
Cisco Kid	Syndicated (Blair Ent.)
Double Dare	Syndicated (Viacom)
Dr. Fad	Syndicated (Fox/Lorber)
Dungeons and Dragons	Network (CBS)
Fun House	Syndicated (Warner Bros.)
Littlest Hobo	Syndicated (Warner Bros.)
Muppets	Syndicated (ITC)
Pee-Wee's Playhouse	Network (CBS)
Peppermint Place	Syndicated (Electra Pictures)
Saved by the Bell	Network (NBC)
Superman	Syndicated (Warner Bros.)
Young Universe	Syndicated (Behrens)

First-Run Weekly Half-Hours

Program	Distributor
Krypton Factor	Syndicated (Samuel Goldwyn)
Lone Ranger(new)	Syndicated (Palladium)
Munsters	Syndicated (MCA)
Sgt. Preston of the Yukon (new)	Syndicated (Palladium)

Notes

1. Wilbur Schramm, Jack Lyle and Edwin B. Parker, *Television in the Lives of Our Children* (Stanford, CA: Stanford University Press, 1961), pp. 13–15.
2. For more recent studies on television and its effects on children, the following sources would be most helpful: 1) Robert Liebert and Emily Davidson, *The Early Window: Effects of Television on Children and Youth* (New York: Pergamon Press, 1973); 2) Rose Goldsen, *The Show and Tell Machine* (New York: Dell Publishing Company, 1979); 3) Peggy Charren and Martin W. Sandler, *Changing Channels*: Living *(Sensibly) with Television* (Reading, Ma: Addison-Wesley Publishing Company, 1983); and 4) Patricia Marks Greenfield, *Mind and Media* (Cambridge, MA: Harvard University Press, 1984).
3. Tom Engelhardt, "The Shortcake Strategy," in *Watching Television*, Todd Gitlin, ed. (New York: Pantheon Books, 1986), p. 71.
4. *Ibid.*
5. "Ban Against Advertising in Children's Programs?" *Broadcasting* (November 20, 1970), p. 56.
6. At the time of this discussion, television and advertising executives quoted a Harvard Business School study, which questioned children's beliefs about what they were told on Saturday morning TV ads. The researchers found that as early as second grade, children indicated a "concrete distrust of commercials, often based on experience with

cials and 'tricky elements' of commercials." By sixth grade, they showed a "global distrust" of all commercials except public service announcements (Joseph Morgenstern, "Children's Hour," *Newsweek* [August 16, 1971], p. 9).

7. Engelhardt, p. 71.

8. Harry Waters, "The Ms. Fixit of Kidvid," *Newsweek* (May 30, 1988), p. 69.

9. Ray Robinson, "Big Bird's Mother Hen," *50 Plus* (December 1987), p. 26.

10. William Melody, *Children's Television: The Economics of Exploitation* (New Haven and London: Yale University Press, 1973), p. 39.

11. Castleman and Podrazik, p. 94.

12. *Ibid.*, p. 95.

13. Melody, p. 40.

14. Ultimately, Walt Disney Productions moved to NBC; but special credit should be accorded to ABC for originating the innovative idea of the film company-network cooperative venture.

15. Castleman and Podrazik, pp. 143–144.

16. "The Bugs Bunny Show," when compared to "The Flintstones," was arguably the better of the two prime-time animated series aired in 1960-1961. Ironically, however, it lasted only 2 years on a weeknight schedule before being relegated to Saturday mornings.

17. "Television for Children: There's More Than May Meet the Eye," *Broadcasting* (November 20, 1972), pp. 31-46.

18. *TV Guide Almanac*, Craig T. and Peter G. Norback, eds. (New York: Ballantine Books, 1980), p. 402.

19. Engelhardt, pp. 95-96.

20. "Children's Television and ACT," "Chronicle," WCVB-TV (April 13, 1988).

21. This phenomenon is referred to as the "halo effect"—a loss in TV revenues due to the lack of popular toys, ad agencies, and retailers.

22. Michael Burgi, "Can Indies Afford to Have Children," *Channels* (January 1989), p. 75.

23. David S. Wilson, "Tooned On! America Flips for Cartoons," *TV Guide* (June 9, 1990), p. 24.

24. "TV Viewing Top After-School Activity for Children," *Broadcasting* (January 11, 1988), p. 56.

25. *Ibid.*

26. *Kiddult* indicates a mixture of both child and adult viewing appeal.

27. While these program genres are clearly used during some daytime "children's hours," they will not be indicated in the show list provided for this chapter. Instead, please consult Chapter 6—"Specialized Fare."

28. People meters are a new controversial method of ratings measurement. It involves more active viewer participation as well as highly specialized computer control boxes.

29. Wayne Friedman, "Kid's Television Under Siege," *Adweek* (January 23, 1989), p. 1.

30. Burgi, pp. 74–75.

31. Wilson, p. 23.

32. *Ibid.*, p. 22.

33. The programs listed on these pages are specifically identified as *children's shows*. Thus, while certain off-network productions and first-run syndications are often scheduled during prescribed kidvid hours, they also appeal to other target audiences. These Show lists may be found in Chapter 6, "Specialized Fare."

6

▼ Specialized Fare

Unlike the other programming genres discussed in this book, Chapter 6 will address show formats that are targeted either to large, hybrid groups, or to smaller, but very specialized audiences during particular time blocks of the daytime week. More specifically, the following sections will describe popular program forms for 1) kiddults (off-network reruns, movie packages, and reality shows)[1]; 2) teenagers and young adults (music and music information); and 3) adult male audiences (sports and sports information). One point must be made very clearly at the outset, though: the demographic distinctions identified in these programming breakdowns are not exclusive; they simply define the primary target audience. For example, obviously, TV sports events are not viewed solely by men. Adult male viewers, however, are not usually recognized as being "available" during most daytime hours. As a result, networks (and stations) try to direct specific programming to this demographic cluster when it is most likely to be accessible—usually during weekend afternoons.

Family-oriented shows and music programs also carry disclaimers similar to the one mentioned above. They each attract very specific primary audiences, but acknowledge other possible viewer groups as well.

OFF-NETWORK RERUNS

Generally, broadcast programmers divide all off-network reruns into two basic categories: vintage shows (otherwise known as *classics* or *evergreens*), and more recent network fare. Evergreens are distinguished from other forms of off-network programming by the following characteristics:

- They include productions from the 1950s, 1960s, and early- and mid-70s.
- For the most part, they have never been cancelled by a network. Instead, the stars (or producers, or both,) simply decided to move on to other ventures.
- They seem to have a "perpetual ability to draw viewers of all ages, children and parents enjoying programs with familiar names and familiar story lines."[2]
- Their plot lines are timeless; devoid of contemporary political or sociological commentary, or both.
- Their costs are relatively low,[3] yet their potential returns are high. Thus, they are considered to be an ideal type of cost-efficient programming.[4]

On the other hand, more recent off-network syndication offerings are usually described as follows:

- Often, they are faster-paced and more technologically slick than vintage shows. They have also been filmed (or taped) in color—a factor which may or may not apply to evergreens.
- Because these shows have been produced from the late 1970s to the present, they usually reflect a political or sociological context.
- They generally contain highly recognizable celebrities, some of whom actually have gained their fame from the shows' original network exposure.
- Unlike vintage programming, some of these shows may have ultimately been cancelled by their networks. However, at some point in time, they enjoyed a certain measure of popularity, and thus, were assigned to relatively long runs on the prime-time schedule.
- Because of increasingly complex above- and below-line production costs, the syndication fees for these programs are often much higher than for classic shows.

Whether it be vintage programming or shows from a more recent decade, the greatest advantage of off-network programming is its appeal to several demographic viewer groups. In addition, reruns can cost as much or as little as the TV manager wishes, thereby providing welcome economic relief in the broadcast schedule. This factor is especially important to non-affiliated station owners, who often must find a balance between competitive programming and small budget allotments. Carol Martz, a program director for KCOP (Los Angeles), says:

> It's very difficult for an independent to have an unproven show [as a success] in the daytime. There is so little money generated in daytime in this market, so everybody sells very similarly in daytime rotation. It doesn't make a lot of sense to put a lot of money into daytime. So it's profitable to stay with this kind of programming.[5]

However, despite general agreement among broadcasters that off-network re-runs are both desirable and necessary for daytime survival, their programming choices and strategies often differ greatly.

One example of successful off-network programming is the family-oriented station, WFFT-TV, in Fort Wayne, Indiana. Jeff Evans, general manager at WFFT, favors vintage fare for his station. He says the following:

> These shows are what people grew up with. "Andy Griffith," "Leave It To Beaver," and "The Brady Bunch" are better than a hell of a lot of the crap on the network or coming off today.[6]

Gail Brekke, general manager at KITN-TV (Minneapolis), concurs with Evans' rather pithy assessment, and adds:

> When you see big specials like "Return To Gilligan's Island," "Return To Mayberry," and the "New Leave It To Beaver" series, it just tells you there's something about these shows. They're classic Americana, and we like these characters.[7]

Like Evans and Brekke, many station managers (both independent and network-affiliated) strongly support evergreen programming because of its obvious financial

1991 Off-Network Availabilities
(Based on 4-Year Network Play)

Program	Distributor
A-Team	MCA
A Different World	Carsey/Werner
Air Wolf	MCA
Alice	Warner Bros.
Alf	Warner Bros.
All in the Family	Viacom
Amen	MCA
Andy Griffith	Viacom
Angie	Paramount
Archie Bunker's Place	Columbia
Avengers	Orion
B.J./Lobo	MCA
Barnaby Jones	Worldvision
Barney Miller	Columbia
Batman	20th Century Fox
Beauty and the Beast	Witt/Thomas
Benson	Columbia
Beverly Hillbillies	Viacom
Bewitched	DFS
Black Sheep Squadron	MCA
Blue Knight	Warner Bros.
Bob Newhart	Viacom
Bonanza	Republic
Bosom Buddies	Paramount
Brady Bunch	DFS
Branded	King World
Buck Rogers	MCA
Chips	MGM/UA
Cagney and Lacey	Orion
Cannon	Viacom
Car 54, Where Are You?	Republic
Carol Burnett	C.B. Distribution
Carson Classics	Columbia
Charlie's Angels	Columbia
Cheers	Paramount
Cosby Show	Viacom
Crazy Like a Fox	Lexington
Dennis the Menace (colorized)	Qintex
Designing Women	Columbia
Dick Van Dyke Show	Viacom
Diff'rent Strokes	Columbia
Duet/Open House	UBU Prods.

1991 Off-Network Availabilities (continued)
(Based on 4-Year Network Play)

Program	Distributor
Dukes of Hazzard	Warner Bros.
Eight is Enough	Warner Bros.
Facts of Life	Columbia
Fall Guy	20th Century Fox
Fame, Fortune and Romance	TPE
Family Affair	Viacom
Family Ties	Paramount
Fantasy Island	Columbia
Flying Nun	Columbia
Full House	Warner Bros.
Get Smart	Republic
Gidget	Lexington
Gilligan's Island	Turner Program Sales
Gimme a Break	MCA
Golden Girls	Buena Vista
Gomer Pyle	Viacom
Good Times	Columbia
Growing Pains	Warner Bros.
Guns of Will Sonnett	King World
Gunsmoke	Viacom
Happy Days	Paramount
Hardcastle and McCormack	Lexington
Hart to Hart	Columbia
Hawaii 5-0	Viacom
Head of the Class	Warner Bros.
Here's Lucy	Warner Bros.
High Chaparral	Republic
Highway to Heaven	Genesis Ent.
Hill St. Blues	Victory
Hitchcock Hour	MCA
Hogan Family	Warner Bros.
Hogan's Heroes	Viacom
Honeymoomers	Viacom
Hunter	Tele-ventures
I Dream of Jeannie	DFS
I Love Lucy	Viacom
I Married Joan	Weiss Global
Incredible Hulk	MCA
It's Gary Shandling	Viacom
Jake and the Fatman	Viacom
Jeffersons	Columbia
Joey Bishop Show	Weiss Global

1991 Off-Network Availabilities (continued)
(Based on 4-Year Network Play)

Program	Distributor
Kate and Allie	MCA
Knight Rider	MCA
Kojak	MCA
Laugh-In	Warner Bros.
Laverne and Shirley	DFS
Leave It to Beaver	Paramount
Life of Riley	New World
Little House On the Prairie	Worldvision
Lost in Space	20th Century Fox
Love Boat I	Worldvision
Love Boat II	Worldvision
M*A*S*H	20th Century Fox
MacGyver	Paramount
Magnum, P.I.	MCA
Make Room for Daddy	Weiss Global
Manniz	Paramount
Married With Children	Columbia
Matlock	Viacom
Matt Houston	Warner Bros.
Maude	Columbia
Mayberry R.F.D.	Warner Bros.
McHales's Navy (colorized)	Qintex
Monkees	Lexington
Mission: Impossible (original)	Paramount
Mork and Mindy	DFS
Mr. Belvedere	20th Century Fox
My Favorite Martian	Warner Bros.
My Little Margie	Weiss Global
My Two Dads	Columbia
My World and Welcome To IT (remastered series)	Republic
Newhart	MTM
Night Court	Warner Bros.
Night Gallery	Warner Bros.
9 To 5	20th Century Fox
Odd Couple	DFS
One Day At a Time	Columbia
Partridge Family	DFS
Perfect Strangers	Warner Bros.
Perry Mason	Viacom
Police Story	Columbia
Police Woman	Columbia

1991 Off-Network Availabilities (continued)
(Based on 4-Year Network Play)

Program	Distributor
Remington Steele	MTM
Soap	Columbia
Square Pegs	Columbia
Star Trek	Paramount
Streets of San Francisco	Worldvision
T.J. Hooker	Columbia
Tales of the Texas Rangers	Columbia
Taxi	Paramount
That Girl	Worldvision
That's My Mama	Columbia
The Man From U.N.C.L.E.	Turner Program Services
The Prisoner	ITC
The Ropers	Taffner
Three's Company	Taffner
Too Close for Comfort	Taffner
Topper	King World
Tour of Duty	New World
Tracey Ullman	Fox
Trapper John	20th Century Fox
12 O'Clock High	20th Century Fox
Twilight Zone	MGM/UA
227	Columbia
Vegas	20th Century Fox
Voyage to the Bottom of the Sea	20th Century Fox
Waltons	Warner Bros.
We Love Lucy	Viacom
Webster	Paramount
What's Happenin'?	Lexington
Who's the Boss?	Columbia
WKRP in Cincinnati	Victory
Wonder Woman	Warner Bros.
Wonderful World of Disney	Buena Vista
Wyatt Earp	Columbia

**1992 Projected Off-Network Availabilities
(Based on 4-Year Network Play)**

Program	Distributor
Anything But Love	20th Century Fox
China Beach	Warner Bros.
Dear John	Paramount
Empty Nest	Witt/Thomas/Harris
In the Heat of the Night	MGM/UA
Just the Ten of Us	Warner Bros.
Leave It to Beaver (colorized)	Qintex
Midnight Caller	Lorimar TV
Mission: Impossible	Paramount
Murphy Brown	Warner Bros.
Paradise	Lorimar TV
Roseanne	Carsey-Werner
Wonder Years	New World

**1993 Projected Off-Network Availabilities
(Based on 4-Year Network Play)**

Program	Distributor
Alien Nation	Johnson Prods.
Baywatch	GTG Ent.
Booker	Stephen Cannell
Coach	MCA
Doogie Howser, M.D.	Steven Bochco/20th Century Fox
Family Matters	Lorimar/Telepictures
Famous Teddy Z	Columbia
Hardball	Columbia
Life Goes On	Warner Bros.
Major Dad	MCA/Universal
Mancuso, F.B.I.	NBC Prods.
Quantum Leap	MCA
Wolf	CBS Entertainment
Young Riders	MGM/UA

reward and viewer appeal. They contend that vintage TV fare possesses only one glaring disadvantage: its black-and-white format. However, this factor may actually be less important than most people may think. Research shows that the writing and acting quality of these programs often supercedes any inferiority in production value.[8] Even younger viewers seem oblivious to older show formats; in fact, they sometimes actually believe evergreens to be as new as everything else on the screen. A small child once remarked:

> I really love "Andy Of Mayberry" [the syndicated title of the old "Andy Griffith Show"]. It's so much fun to watch. Maybe next year, they'll make it in color. Do you think so?[9]

Obviously, this child could detect little difference between excellence in first-run programming versus that in classic reruns.

Recent off-network programming is much more expensive and, hence, much riskier than that of the evergreen variety. Consequently, most station managers expect some type of success profile for the shows they may wish to buy. Several syndication companies have already begun to provide this data in various packages. Most notable among the latest innovators is Warner Bros. Domestic TV Distribution. Journalist Brian Lowry has described the Warner Bros. *syndication success formula* as an extension of Paramount TV's original *syndicator indicator*. Both systems work as follows:

> [They] deal with [the] demographic composition of a show's network audience....[According to the WB formula] "hit" shows in syndication had a higher percentage of young adult viewers, with 47% or more of its audience age 18-49, and teen/kid viewers (29% or more). They also have a lower percentage of viewers 50 and older (24% or less) and greater male appeal. "Misses" have fewer young adults (40% or less), fewer teens/kids (22% or below) and more viewers in the 50-plus group (39% and up), who aren't likely to be following the show into syndication. Some of the past hits that meet [the former] criteria include "M.A.S.H.," "Three's Company," "Happy Days" and "Barney Miller," while shows that tended to have an older audience on the network such as "The Mary Tyler Moore Show," "All in the Family," and "Webster"—also major hits at the network level—proved disappointments in syndication.[10]

However, despite syndicators' attempts to identify potentially successful off-network reruns, most station managers prefer the proven track records of vintage shows for most of their daytime slots. The only exception to this rule seems to be in the weekday early fringe hours (4:00–6:00 P.M.), when the teen/child and young adult audiences are larger than in other parts of the day.[11]

MOVIE PACKAGES

Movie distribution, like off-network syndication, is comprised of two major program types: vintage productions, and recent theatrical releases. Today, barter movie packages seem to be a popular trend among TV programmers;[12] and they are presented at all hours of the broadcast day. Most station executives, however, prefer to air vintage movies in daytime. Their reasons are clear: old movies (like evergreen shows) cost comparatively little, and yield lucrative revenues. In addition, local programmers often frame their movie fare around local hosts or hostesses, thus creating a stronger station affiliation with its community.

REALITY SHOWS

Despite the recent programming surge in emergency rescues, medical diagnoses, and armchair adjudications, the concept of *reality-oriented programming* is not new. The first attempt at dramatizing reality occurred in the early 1950s, with the introduction of legal sociodrama. Among the most popular shows in this era was "They Stand Accused," a product of the DuMont network. "Accused" became the prototype for many courtroom dramas at this time, since it was based on a simple concept and was very inexpensive to produce. However, after the introduction of "Perry Mason" by CBS in 1957, audiences came to expect dramatic programming with greater aesthetic value. As a result, many shows like "Accused" disappeared from the prime-time network rosters;[13] but they continued to prosper in daytime and on local station schedules. In fact, some of the more popular reality shows throughout the 60s and into the 70s originated in local studios and were later syndicated.

Reality Shows—1991 Availabilities
Current First-Run Programs

Program	Distributor
Cop Talk—Behind the Shield	Tribune Ent.
Crime Stopper 800	All American
Divorce Court	Blair Ent.
Missing/Reward	Group W
Only Yesterday	King World
People's Court	Warner Bros.
Secrets and Mysteries	ITC
Tales of the Unexpected	Orbis
The Judge	Genesis
The Making of ...	Muller Media
Witness to Survivial	S.F.M.

Current/Future Off-Network Reruns

Program	Distributor
Jacques Cousteau	Turner Program Sales
Rescue 911	CBS Entertainment
Million Dollar Video Challenge	World Events
That's Incredible	MCA
Totally Hidden Video	Quantum Media
Unsolved Mysteries	Cosgrove/ Meurer

They included programs such as "Traffic Court," "Divorce Court," and "Day In Court."[14]

By the mid-70s, viewers had seemingly grown tired of the reality format. They turned, instead, to game shows, talk/variety offerings, and local programming for their daytime entertainment. As a result, many media analysts predicted the ultimate demise of reality-oriented shows. While it appeared for a time that the critics were correct in their assessment, they were later proven to be wrong when the smash syndication hit, "The People's Court," appeared on TV schedules.

The setting for "People's Court" was a Los Angeles sound stage, where actual litigants argued their cases in a simulated small-claims courtroom. Unlike other legal dramas, however, the plaintiffs, the defendants, and their points of contention were often common and petty. Author Michael Pollan notes:

> There are no production values whatsoever—each show boasts a single, standard-issue courtroom set; "action" is when someone rises to testify, and the nearest thing to props are the defective camshafts and dog-soiled carpets that the plaintiffs on "People's Court" enter into evidence. Imagine a show that combines the production values of C-SPAN with the drama of a shopping channel and you have some idea of the level of excitement these programs reach.[15]

Nevertheless, the ratings climbed. In 1987, "People's Court" was found to be the fourth-most-popular program in syndication. It still ranks as one of the most highly-rated first-run shows today.

"People's Court" celebrity judge, Judge Joseph A. Wapner, seems to think that the program's huge success stems from its basic simplicity and realism. Judge Wapner notes the following:

> . . . I have never lost sight of the significance and value of small-claims court. Other than traffic court, small-claims court is the only contact most Americans have with our judicial system. Although cases are typically resolved in a short period of time—usually fifteen minutes—that's all it takes for litigants to form their impressions of our judicial system. Win or lose, people must feel they had their day in court; they must believe the judge listened to their side of the story and gave the matter serious consideration before rendering a decision.

> . . . Although most cases involve relatively modest sums of money, they often have an important impact on the lives of the litigants. The woman who saved for two years to buy a used car that fell apart, the man whose neighbor demolished their common wall, the customer whose best shirt was ruined by the dry cleaner, all care a great deal about vindicating their position and upholding their rights. Oddly enough, in many cases, individuals with fifty-dollar disputes over faulty products are aroused more than the company president who is sued for $50 million for antitrust violations...the amount of money involved in a dispute often bears no resemblance to the importance of the case.[16]

Whatever the reason for its popularity, the positive viewer response to "People's Court" quite naturally has led more producers to create quasi-realistic legal, medical, and law enforcement imitators. Some of these shows (like "Group One Medical," "Family Medical Center," and "On Trial") have barely made it past their first seasons. However, other programs, such as "The Judge" and a revised "Divorce Court" have done quite well in the ratings.

Some media critics relate a specific reality show's success quotient to its subject matter, that is, medical dramas are said to be less popular because talk show celebrities like Oprah Winfrey and Phil Donahue often address similar topics in a more appealing manner.[17] However, most of the time it can be said that a program's overall format may be the key factor to its ultimate rise or fall from viewer affinity. The following examples, are noted by Michael Pollan:

- "Divorce Court" is a soap opera derivative. Based on the innumerable cases of adultery, deceit, substance abuse, and corruption, one must conclude that its target audience is comprised mainly of those viewers who "watch the usual crew of daytime adulterers, backstabbers, childnappers, wife abusers and philanderers finally get their comeuppance."[18] Incidentally, many of the actors who portray the litigants on "Divorce Court" have actually either been on soap operas previously or hope to become castmembers in the future.
- "The Judge" attempts to deal with topics found on talk shows like "Sally Jessy Raphael," "The Oprah Winfrey Show," "Donahue" and "Geraldo". Past cases have included such controversial issues as surrogate motherhood, product liability, sex discrimination, drunken driving, and book banning. Despite its single-minded goal of introducing current issues to the viewing public, however, the overall response to the program has been comparatively unenthusiastic. Pollan surmises that one of the reasons for this lack of viewer interest is due to "one of the anomalies of courtroom shows—their relative popularity seems to be in inverse relation to their dramatic intensity."[19]
- "People's Court" seems to follow the game show format most closely. Viewers are able to side with the people they like, talk back to their television sets, and test their logic according to Judge Wapner's legal expertise. Further, this format "feeds our fantasies of revenge against the dry cleaners, used car dealers and neighbors whose abuse we suffer every day."[20]

Based on the track records of these shows and of other reality-based programming, this genre seems to be healthier that it has ever been in TV history. Thus, barring the possibility of overexposure, reality shows are likely to be a daytime broadcast staple for several years to come.

MUSIC PROGRAMS

Music-oriented TV shows were introduced into network daytime in 1950, when Kate Smith hosted NBC's first afternoon variety program.[21] At this time, most televised music shows were like their radio counterparts; they consisted of popular mainstream songs, intended for nonsegmented audiences. Just a few years later, however, both radio and TV programmers recognized the value of formatting different songs for various demographic groups. Most especially, they witnessed a new musical trend (called "rock 'n' roll"), which created an entirely new target market of teenagers and young adults—the nation's primary record buyers as well as a generally attractive consumer population.

By 1956, many former middle-of-the-road pop singers had faded into oblivion due to the new craze of rock 'n' roll, which, by now, had become firmly entrenched in most American media. Relatedly, daytime television began to seek show formats that would capitalize upon the latest musical mood of the country. Several network concepts for teen-oriented programming were introduced at this time, but most of these ideas failed. Ironically, the only unqualified music hit during this era was a locally-produced show that had been aired by its network only after it had proven to be successful for five years in Philadelphia. This rather unpretentious dark horse was named "American Bandstand," which was introduced by ABC to a nationwide audience in August 1957. The format of "Bandstand" was very simple, but extremely effective; and despite dire predictions by the show's detractors, it soon became a TV staple for the nation's younger viewing audience.

The key to the success of "American Bandstand" was its producer and disc jockey, Dick Clark, who was a businessman as well as a rock 'n' roll fan:

> He realized that the nature of a live rock performance did not transfer easily to the small, confining TV screen. While masters at their music, most rock performers were novices at projecting any visual, physical stage presence (beyond shaking) for television. Clark provided the necessary stabilizing control and guidance. He also recognized the need to place rock 'n' roll in safer, more accessible surroundings for general consumption. On his programs, Clark was always neatly dressed, clean-cut, warm, and articulate. He usually emphasized less threatening personalities such as Fabian, who did toned-down versions of black rhythm and blues hits.[22]

"American Bandstand" quickly caught on with most American teenagers, and it became an unqualified success for ABC in its first year of network exposure. After several years of popularity in its weekday early fringe slot, "Bandstand" moved to prime-time television for a brief stint. It later was reassigned to Saturday afternoons, where it competed against several of its imitators for many years.

In the mid-70s, the music/variety/dance genre seemed to waver in viewership, however. Some shows survived, but production companies were clearly concentrat-

Music Shows—1991 Availability	
Program	*Distributor*
Big Break	Multimedia
Classic Country	Genesis Ent.
Dionne and Friends	Tribune Ent.
Entertainment Coast to Coast	Viacom
Fabian Presents	Omnivision
Fabian Turns It Loose	Omnivision
Hee Haw	Gaylord
Music City, U.S.A.	Multimedia
Showtime at the Apollo	Raymond Horn
Soul Train	Tribune Ent.
Star Search	TPE

ing most of their efforts on game shows and talk shows. This mood prevailed until 1981, when cable television introduced MTV, a 24-hour music network that changed the face of music presentation.

MTV became an overnight sensation by capitalizing upon the new technology of music video production. Viewers were now able to see their favorite songs performed on the screen. They also were treated to celebrity interviews, music news, and quirky game shows, courtesy of several popular VJs (video jockeys). In addition, MTV breathed new life into the radio, television, and recording industries by proving that a new approach to an old idea can be very profitable. Author Fredric Dannen notes:

> MTV revitalized the record industry by giving play to acts radio ignored. Its influence led movie-makers to long-form videos such as "Flashdance" and "Purple Rain." The working title for "Miami Vice" was "MTV Cops."[23]

Since its debut, MTV has suffered some typical growing pains. In the mid-80s, the network was accused of being racist in its video airing decisions. Also, during that time, some of MTV's original fans claimed that its reputation for presenting "cutting edge" material had eroded. Despite these problems, however, no one can deny the great impact MTV has had on broadcast and cable programming.

Today, music shows are produced for most segments of the viewing and listening population. While the ratings are not overwhelming in this category, the fact remains that music programming has a solid place in television.

SPORTS

Many sports programs or events, or both, are broadcast on Saturday and Sunday afternoons. Because of this the sports genre has been included in this chapter. It should also be noted that sports has always been a vital part of television. In 1948, for example, most network programming (over 60%) was devoted to football, baseball, wrestling, or boxing.

Sports Shows—1991 Availability

Program	Distributor
American Gladiators	Syndicated (Samuel Goldwyn)
Driver's Seat	Syndicated (MCA)
In Sport	Syndicated (Select Media)
Motorweek Illustrated	Syndicated (Orbis)
Roller Games	Syndicated (Qintex)
Ski Scene	Syndicated (Raymond Horn)
Sports Sunday	Network (CBS)
Sportsworld	Network (NBC)
Tuff Trax	Syndicated (Qintex)
Water Sports World	Syndicated (Great Ent.)
Wide World of Sports	Network (ABC)

The programs found in the 1991 show list reflect a rather recent diversity in TV sports offerings. Station managers have discovered many of these areas of viewer interest due to the presence of sports cable networks. As audiences continue to mobilize and segment, programming in this genre is likely to expand and change further. The direction is unknown at this point, however.

Notes

1. As discussed in an earlier chapter, *kiddults* are audiences made up of both children and adults.
2. Eliot Tiegel, "Vintage Shows: All Legs, No Cash," *Television/Radio Age* (November 14, 1988), p. 47.
3. According to Tiegel (see previous footnote for cite), vintage shows can be acquired for as low as $50 per episode in small markets. Larger ADIs can obtain them for just a few hundred dollars a program.
4. Tiegel, p. 47.
5. *Ibid.*, p. 49.
6. *Ibid.*, p. 47.
7. *Ibid.*, p. 49.
8. *Ibid.*
9. This comment was made by the author's four year-old niece, Lisa, in 1984.
10. Brian Lowry, "WB's Formula for Off-Net Sitcom Success Gets Mixed Reaction from Station Execs," *Variety* (November 28, 1989), p. 51.
11. As a result, most recent off-network fare is generally programmed during access periods, prime time, and late fringe hours.
12. Robert Sobel, "Big Glut of Barter Movie Packages; Impact on Syndication Sales Unclear," *Television/Radio Age* (September 5, 1988), pp. 53-54. Some of the syndication companies most active in barter movie packages include Televentures, Qintex, Orion Television, Orbis, Multimedia, Viacom, and Columbia Pictures Television.
13. Castleman and Podrazik, p. 122.
14. *Ibid.*, p. 128.
15. Michael Pollan, "Reality Shows: The Syndicated Bench," *Channels* (July/August 1987), p. 52.
16. Harvey Levin, *The People's Court: How to Tell It to the Judge* (New York: Quill, 1985), p. 12.
17. Robert Sobel, "'Oprah' Have-Nots Look for Answers in Early Fringe," *Television/Radio Age* (December 12, 1988), p.44.
18. Pollan, p. 53.
19. *Ibid.*
20. *Ibid.*, p. 54.
21. "The Kate Smith Show" premiered September 25, 1950.
22. Castleman and Podrazik, pp. 121-122.
23. Fredric Dannen, "MTV's Great Leap Backward," *Channels* (July/August 1987), p. 45.

7

▼
▼
▼
▼
▼

Putting It All Together

After reviewing all of the possible show genres available for schedule selection, it is apparent that a successful TV programmer must possess: (1) the ability to identify a profitable audience for each time slot; (2) the skill needed to disburse the station's budget in a responsible, yet imaginative manner; (3) the creative acuity to balance enough locally produced programming with syndicated (or network) fare to generate a strong station image; and (4) the competitive edge to handle growing numbers of both broadcast and cable outlets.

The talent needed to meet these high standards for station success become even more challenging during the daytime hours, when broadcasters are given (comparatively) little money; and have smaller audiences, who may or may not be paying much attention to the carefully crafted programming strategies found on their television screens.

THE AUDIENCES

To better understand today's daytime viewing audience, a random survey of over 600 persons was taken in 1985.[1] Subjects were asked how much television they watched, which daytime programs they preferred, and why they chose to watch these programs. Table 5 illustrates the overall results of these interviews.

In general, most subjects who had a free afternoon often listed watching TV as one of their primary choices. However, different age groups voiced different priorities for their various viewing habits. For example, children between the ages of 1 and 10 often substituted daytime television for some outside activity or getting their homework finished (so they could watch TV later). An overwhelming number of 11 to 20 year-old females chose television as their primary recreation; and many of the male college-aged students were also heavy viewers.

However, as the 11 to 20 year-olds moved into the 21 to 40 age category (and left school), an interesting change occurred. Men continued to watch the same amounts of daytime television, but women declined in their viewing amounts. When interviewed, these females commented that their change in marital status, professional commitment, and in having children and maintaining a household accounted for their changes in habit. However, an impressionable number of females aged 21 to 40 said they still watched soap operas occasionally; and many times videotaped them during the day so they could view them at their convenience.

▶ **Table 5**   Preferences for daytime activities

Gender/Age	% who watch TV	% doing work in.	% going to films	% doing an out. act.	% doing other
Totals: 1-10(24)	29.2%	—	—	45.8	33.3
Males 1-10(12)	41.7	—	—	50	25
Females 1-10 (12)	16.7	—	—	41.7	41.7
Totals: 11-20 (187)	77	12.3	2.1	25.7	19.8
Males 11-20 (59)	59.3	15.2	1.7	35.6	25.4
Females 11-20 (128)	85.2	10.9	2.3	21.1	17.2
Totals: 21-40 (249)	69.5	26.9	4	26.1	14.9
Males 21-40 (93)	59.1	18.3	5.4	35.5	12.9
Females 21-40 (156)	75.6	32.1	3.2	20.5	16
Totals: 41-60 (99)	47.5	46.5	—	31.3	10.1
Males 41-60 (36)	30.6	36.1	—	52.8	11.1
Females 41-60 (63)	57.1	52.4	—	19	9.5
Totals: 61+ (42)	54.8	35.8	2.4	19	28.6
Males 61+ (16)	50	31.2	6.3	25	25
Females 61+ (26)	57.7	38.5	—	15.4	30.8

NOTE: The numbers in parentheses indicate the actual numbers of persons surveyed in each age/gender group.

Between ages 41 to 60, total viewing was less than in the previous two age groups. Both men and women preferred to use their free time for shopping, errands, lunch with friends, and so forth. Not surprisingly, TV viewing dramatically increased for persons 61 and over.[2]

Delving further into the analysis, Tables 6 and 7 reflect the priorities of those persons who claimed to spend most of their time watching television during the day. Table 6 presents overall programming preferences, and Table 7 provides a gender/age breakdown of the numbers.

Based on the informations from these two tables, the following generalizations can be made:

1. *Ages 1 to 10*. Not surprisingly, both girls and boys chose cartoons as their favorite type of daytime programming. However, girls were evenly split in making their second choice: they found soap operas and game shows to be equally enticing; boys, on the other hand, showed no strong preference for anything outside of cartoons.

2. *Ages 11 to 20*. Men and women in this age group watched all categories of programming (soap operas, game shows, movies, talk shows, sports, cartoons, and old reruns), but soap operas were by far the most watched. The most popular second choice for men seemed to be cartoons; women were almost totally devoted to soaps, but if forced to make another selection, chose game shows. It is important to note that many people in this age category were college students; and this evidence serves to support other research claiming that college students are the second largest soap opera audience in television.

3. *Ages 21 to 40.* Members of both sexes in this category also overwhelmingly chose serial drama as their favorite form of daytime TV programming. However, men in this age group were more open to second- and third-choice categories (cartoons, old reruns) than their female counterparts. Women were almost totally committed to soap operas (91.4%). Their next preference, game shows, drew only 14.4% of the interviewees.

4. *Ages 41 to 60.* Again, soap operas were extremely popular with this group (74.5%); and movies were chosen as a distant second (21.3%). However, males in this age range were the first adult group not to select soap operas as their first choice. Instead, most preferred movies as a 2:1 favorite over soaps and game shows. Two other surprising observations should be noted. First, it seemed amazing that a fairly large percentage of men in this age range watched soap operas at all—they were not identified initially as a soap-oriented group. Secondly, a fairly large percentage

▶ **Table 6** Program preferences of daytime television audiences

Program	Percentage of Preference
Soap Operas	77.7%
Game Shows	16.8%
Movies	11.9%
Cartoons	11.7%
Other (News, PBS, Sports, Reruns, Talk Shows, First-Run Syndication)	11.4%

▶ **Table 7** Viewers' choices of viewing

Gender/Age	Soap Operas	Game Shows	Movies	Cartoons	Other*
Totals 1-10 (7)	42.9%	42.9%	14.3%	100%	14.3%
Males 1-10 (5)	40%	40%	20%	100%	20%
Females 1-10 (2)	50%	50%	—	100%	—
Totals 11-20 (144)	82.6%	14.6%	5.6%	12.5%	11.1%
Males 11-20 (35)	51.4%	20%	14.3%	34.3%	17.1%
Females 11-20 (109)	92.7%	12.8%	2.8%	5.5%	9.2%
Totals 21-40 (173)	79.2%	15.6%	12.7%	11%	12.7%
Males 21-40 (55)	47.3%	18.2%	21.8%	27.3%	30.9%
Females 21-40 (118)	94.1%	14.4%	8.5%	3.4%	4.2%
Totals 41-60 (47)	74.5%	12.8%	21.3%	4.4%	6.4%
Males 41-60 (11)	27.3%	27.3%	54.5%	18.2%	18.2%
Females 41-60 (36)	88.9%	8.3%	11.1%	—	2.8%
Totals 61+ (22)	54.5%	40.9%	27.3%	—	13.6%
Males 61+ (7)	42.9%	57.1%	28.6%	—	—
Females 61+ (15)	60%	33.3%	26.7%	—	20%

NOTE: The numbers in parentheses indicate the actual numbers of persons who selected watching television as one of their choices. The total percentages will equal more than 100% because many listed 2, 3, or 4 choices.

*"Other" included news, PBS, Sports, MTV, old reruns, talk shows, and first-run syndications.

of men seemed to enjoy cartoons. Apparently, we never outgrow our affection for cartoons.

5. *Ages 61 and over*. Slightly over half of this age group selected soap operas as their first choice. Game shows were chosen next most often. An interesting characteristic of people in this age group was that the viewers were most adamant about their program choices. For example, those who watched soaps usually watched many; those who disliked them were extremely vocal and watched *only* game shows, movies, news, or PBS.

Despite the obvious weaknesses in this 1985 study,[3] several daytime viewing audiences were clearly identified. However, beyond this simple classification, it is difficult, if not impossible, to define a typical daytime audience. This is because several demographic crossovers (such as kiddults) occur during staggered periods of the day (due to differing school schedules, or flextime, and so forth). In addition, the current amount of VCR usage in this country contradicts a previous programming assumption—that only those at home during the day watch daytime television. Thus, successful TV programmers often look at local work schedules, school hours, traffic patterns, and homevideo rental periods to determine the needs of their specific communities.

THE BUDGET

Despite viewer perceptions that stations (and networks) have access to unlimited budgets, TV programmers know better: the key to success is often found in the creative use of restricted monies. This fact is even more imperative during the daytime hours, since syndication fees or local production costs, or both, are generally much greater and more highly prioritized for prime time. As a result, broadcasters must spend their funds wisely, picking and choosing the best times for power shows,[4] counterprogramming,[5] and alternative fare.[6]

To further illustrate this point, it may be helpful to look at some successful show placement stories, as well as some failures, which have occurred in the last several years. In the 1989 daytime television schedule, for example, some interesting audience flow chart data emerged.

- "Jeopardy!" established itself as the most consistently successful news lead-in for network affiliates, delivering an average of 70-75% of its viewers to the news. Second to "Jeopardy!" was "The People's Court," with a news audience delivery average of 60-65%. On the other hand, "The Cosby Show," with its hefty syndication fees, delivered relatively poor lead-in numbers to the news.[7] Thus, one must weigh the syndication cost with its potential return in the time block.
- "The Cosby Show," "Cheers," "M.A.S.H.," and certain issue-oriented talk shows like "Oprah Winfrey" attracted large numbers of children, teens, and young women. This statistic made the programs appear to be very attractive for a 4:00–5:00 P.M. scheduling slot, but not necessarily as lucrative either earlier or later in the day. Obviously, school hours and homework schedules figured prominently in this behavior.[8]

- Reality news magazines, such as "A Current Affair" and "Inside Edition," were relatively inexpensive, and reflected a fairly even demographic mixture of older and younger viewers during both the early fringe and late fringe periods of the broadcast day. Some station managers thought it was a useful lead-in to news,[9] but others may have preferred to separate it from more traditional forms of broadcast journalism.[10]

Another factor related to budgets and show placement is the changing syndication market, which is now demanding more cash only deals instead of the familiar barter negotiations of several years past. The reasons for this change in policy are attributed to the overpopulation of syndicated products—both off-network and first-run—as well as the time slot limitations of potential buyers. Mitchell Praver, Katz Continental vice president, explains the situation in the following way:

> The problem first surfaced with the advent of barter minutes taken out of first-run syndicated programs. This forced the distributors to clear major markets, which in turn forced them into a 70% clearance rate to pull in advertisers. To achieve 70%, oftentimes distributors have to accept clearances that are not the best showcase for new programs. Programs designed for early fringe and access are often buried in early or late night time periods, where they don't get a chance to perform.[11]

As a result, most syndication companies are demanding cash only for their program offerings—a situation which can be extremely expensive and risky for a local TV programmer. It is not surprising, then, that many broadcasters have turned to vintage television shows for major portions of their daytime schedules. KTXA-TV General Manager Dirk Brinkerhoff states:[12]

The audience for this kind of program doesn't seem to go away. Every year a new group of viewers is at home during the day and sees these shows for the first time. They continue to perform, but not at the level of "Bill Cosby" or "Cheers." But they work for the amount of money you pay for them.[13]

A wise TV programmer knows when to buy a show, and for which daypart. Mistakes in either area can cost the station valuable ratings points as well as needed advertiser support.

THE IMAGE AND THE COMPETITIVE EDGE

Whether it be at the network level or at a local station, one of the most challenging dilemmas of television programming has always been the continuous quest for identity within a world of imitation. As broadcasters enter into the twenty-first century, the pressure will hardly abate. In fact, it can only increase, given the growing technology of both CATV and satellite broadcasting.

A TV station's identity usually develops as a result of: (1) its balance of nationally-distributed fare (either network or syndicated) versus locally produced shows; (2) its programmatic theme (*information, comedy, family-oriented*, and so forth); and (3) its overall promotional campaign (both short-term and long-term). From the broadcast grid presented in Figure 2,[14] one is able to see how several stations attempt

MORNING

	6:00	6:30	7:00	7:30	8:00	8:30	9:00	9:30	10:00	10:30	11:00	11:30
2 PBS	OFF–AIR		SESAME ST.		MR. ROGERS	SESAME ST.		MR. ROGERS	W/W OF ANIMALS	READING RAIN–BOW	3-2-1 CON-TACT	SESAME ST.
4 NBC	LOCAL NEWS		TODAY				JOAN RIVERS		LOCAL TALK SHOW		LOVE CONN.	CONCEN-TRATION
5 ABC	LOCAL NEWS		GOOD MORNING AMERICA				LOCAL MORNING SHOW		GERALDO		SALLY JESSY RAPHAEL	
7 CBS	FIRST BUSINESS	CBS NEWS	CBS THIS MORNING				REGIS AND KATHIE LEE		FAMILY FEUD	WHEEL OF FOR.	PRICE IS RIGHT	
25 FOX	OFF–AIR	BOZO'S BIG TOP	CARTOONS				TBA	CHURCH SERV.	700 CLUB		CHIPS	
38 IND	CARTOONS				VINTAGE OFF–NETWORK SITCOMS						DIVORCE COURT	
44 PBS	OFF–AIR				ADVENTURE		AMERICAN MASTERS		W/W OF ANIMALS SUR.	WORLD OF SUR.	MOVIE	
56 IND	HEAD-LINE NEWS		CARTOONS				CHILD.'S ROOM	50 YRS. AGO TODAY	VINTAGE OFF–NET. SITCOMS			
68 IND	BIBLE LESSON	CHILD.'S ROOM	TODAY'S MONITOR		ONE NORWAY STREET		CHILD.'S ROOM		MONITOR FORUM		NATIONAL GEOGRAPHIC	

▶ *Figure 2   A typical broadcast morning in the Boston television market—1990.*

AFTERNOON

	12:00	12:30	1:00	1:30	2:00	2:30	3:00	3:30	4:00	4:30	5:00	5:30
2 PBS	SESAME ST.	THE STORY OF ENGLISH		NATURE		ART SHOW	THIS OLD HOUSE	SESAME ST.		MR. ROGERS	READING RAINBOW	3-2-1 CONTACT
4 NBC	LOCAL NEWS	NBC SOAP OPERAS							3RD DEGREE	FAMILY FEUD	PEOPLE COURT	LOCAL NEWS
5 ABC	LOCAL NEWS	ABC SOAP OPERAS							DONAHUE		OPRAH	
7 CBS	LOCAL NEWS	CBS SOAP OPERAS							INSIDE EDITION	HARD COPY	LOCAL NEWS	
25 FOX	THE JUDGE	TALK-ABOUT	WIN, LOSE, DRAW	VINTAGE OFF-NET. SITCOM	CARTOONS						OFF-NET SITCOM	TRIBES
38 IND	OFF-NET DRAMA		VINTAGE OFF-NET SITCOMS		CARTOONS						FUN HOUSE	NEW GIDGET
44 PBS	MOVIE		CUISINE RAPIDE	MADELEINE COOKS	AMERICAN MASTERS		ADVENTURE		CONSERVING AMERICA		MCLAU. GROUP	COMP. CHRON.
56 IND	OFF-NETWORK SITCOMS				CARTOONS						OFF-NET SITCOMS	
68 IND	TODAY'S MONITOR		ONE NORWAY STREET		NATIONAL GEOGRAPHIC		MONEY AND YOU		INNER CITY BEAT		CHILD.'S ROOM	VINT. OFF-NET.

▶ *Figure 3*   *A typical broadcast afternoon in the Boston television market—1990.*

to find an identity in a major market through national/local balance as well as programmatic themes. Promotional strategies, while not found in these figures, must be adapted to both balance and theme.

As noted from the above broadcast schedule, specific programming strategies may differ; but the overall success of a television operation depends upon its ability to use its budget, image, competitive edge, and creative energies wisely. One final point: the success of a station or network is often measured by the number of imitators who copy its strategies. Thus, a once-innovative scheduling plan must often be discarded because it has become too traditional and staid. That's the greatest weakness of the programming business; it's also its greatest strength.

Notes

1. Matelski, *The Soap Opera Evolution*, pp. 40–42.
2. Actually persons 61 and over were more active in each category than the other age groups. Perhaps this was due to their large amount of discretionary time as compared with younger adults.
3. In all studies dealing with randomly selected subjects, researchers run the risk of creating samples which are not totally representative. In this study, 600 people were randomly selected in several areas of the country. However, because college students served as the data gatherers, it is possible that the resulting sample was somewhat skewed educationally and economically. This does not take away from the fact that today's viewers are educationally and economically better off than in the 40s—other studies have confirmed this trend. However, it is important to note that the percentages may be higher in this analysis than in other research.
4. *Power* programming usually involves a competition between two or more popular shows, that ultimately will split a very large potential audience.
5. *Counterprogramming* is the act of scheduling competitive shows that appeal to certain audiences not otherwise being served. It differs from "power" programming in that it does not assume a overwhelmingly large, multi-segmented audience; however, it does presuppose some moderately high viewer numbers.
6. *Alternative* programming is designed to appeal to a small, highly defined audience that is not likely to be reached by typical mass appeal shows.
7. John Dempsey, "The Leading News Lead-Ins: 'Jeopardy' Tops 'Cosby,'" *Variety* (July 12, 1989), p. 41.
8. *Ibid.*
9. *Ibid.*, p. 42.
10. See "News Magazines," Chapter 2.
11. "Catch 22 Time for Many Syndicators; Want Clearances, Get Unwanted Times," *Television/Radio Age* (October 2, 1989), p. 46.
12. KTXA-TV is located in Dallas, Texas.
13. Eliot Tiegel, "Vintage Shows: All Legs, No Cash," p. 48.
14. This program grid was adapted from the Boston edition of *TV Guide* (June 30, 1990).

BIBLIOGRAPHY

Books

Allen, Robert C. *Speaking of Soap Operas*. Chapel Hill, NC: University of North Carolina Press, 1985.

Busby, Linda J. *Mass Communication in a New Age: A Media Survey*. Glenview, IL: Scott, Foresman and Company, 1988.

Cantor, Muriel and Suzanne Pingree. *The Soap Opera*. Beverly Hills, CA: Sage, 1983.

Castleman, Harry and Walter J. Podrazik. *Watching TV: Four Decades of American Television*. New York: McGraw-Hill Co., 1982.

Charren, Peggy and Martin W. Sandler. *Changing Channels: Living (Sensibly) with Television*. Reading, MA: Addison-Wesley Company, 1983.

Goldsen, Rose. *The Show and Tell Machine*. New York: Dell Publishing Company, 1979.

Graham, Jefferson. *Come On Down! The TV Game Show Book*. New York: Abbeville Press, 1988.

Greenfield, Patricia Marks. *Mind and Media*. Cambridge, MA: Harvard University Press, 1984.

Harless, James D. *Mass Communication: An Introductory Survey, 2nd Ed.* Dubuque, IA: Wm. C. Brown Publishers, 1990.

Head, Sydney W. *World Broadcasting Systems: A Comparative Analysis*. Belmont, CA: Wadsworth Publishing Company, 1985.

Howard, Herbert H. and Michael S. Kievman. *Radio and TV Programming*. Columbus, OH: Grid Publishing, Inc., 1983.

Inglis, Andrew F. *Behind the Tube: A History of Broadcasting Technology and Business*. Boston: Focal Press, 1990.

LaGuardia, Robert. *Soap World*. New York: Arbor House, 1983.

Liebert, Robert and Emily Davidson. *The Early Window: Effects of Television on Children and Youth*. New York: Pergamon Press, 1973.

MacDonald, J. Fred. *Don't Touch that Dial: Radio Programming in American Life from 1920 to 1960*. Chicago: Nelson-Hall, 1979.

Matelski, Marilyn J. *Broadcast Programming and Promotions Worktext*. Boston: Focal Press, 1989.

_____. *The Soap Opera Evolution: America's Enduring Romance with Daytime Drama*. Jefferson, NC: McFarland & Co., Inc., 1988.

Melody, William. *Children's Television: The Economics of Exploitation*. New Haven and London: Yale University Press, 1973.

Radio Research: 1942–1943. Edited by Paul F. Lazarsfeld and Frank N. Stanton. New York: Essential, 1944.

Schramm, Wilbur, Jack Lyle and Edwin B. Parker. *Television in the Lives of Our Children*. Stanford, CA: Stanford University Press, 1961.

Watching Television. Edited by Todd Gitlin. New York: Pantheon Books, 1986.

Willis, Edgar E., and Camillie D'Arienzo. *Writing Scripts for Television, Radio and Film.* New York: Holt, Rinehart and Winston, 1981.

Wyver, John. *The Moving Image: An International History of Film, Television & Video.* New York: Basil Blackwell, Inc., 1989.

Articles

Bierbaum, Tom. "Nets' Nonprimetime Ratings Dive in 1st Quarter of '90." *Variety,* 11 April 1990, p. 79.

Block, Richard C. "History of Syndie, A Short Look Back." *Variety,* 17 February 1988, p. 77.

Burgi, Michael. "Can Indies Afford to Have Children?" *Channels,* January 1989, pp. 74-76.

"Catch 22 Time for Many Syndicators; Want Clearances, Get Unwanted Times." *Television/ Radio Age,* 2 October 1989, p. 46.

Dannen, Fredric. "MTV's Great Leap Backward." *Channels,* July/August 1987, pp. 45–47.

Dempsey, John. "The Leading News Lead-Ins: 'Jeopardy' Tops 'Cosby.'" *Variety,* 12 July 1989, pp. 41–42.

_____. "More Mags Will Fly in Fall; Too Much of a Bad Thing?" *Variety,* 4 April 1989, pp. 79, 98.

_____. "Syndies See '90 as Year of the Gameshow." *Variety,* 13 September 1989, p. 59.

_____. "24 Advertisers Boycott 'Affair,' 'Edition,' 'Copy.'" *Variety,* 21 March 1990, pp. 1, 60.

Friedman, Wayne. "Kids' Television Under Siege." *Adweek,* 23 January 1989, p. 30.

Gelmen, Morrie. "Medical Shows Spearhead Rush to Reality." *Variety,* 17 February 1988, pp. 1, 64.

Haithman, Diane. "Soap Writers Rule Daytime TV with a Godlike Hand." *Detroit Free Press,* 25 November 1984, p. 1G.

Lowry, Brian. "WB's Formula for Off-Net Sitcom Success Gets Mixed Reaction from Station Execs." *Variety,* 29 November 1989, p. 51.

Morgenstern, Joseph. "Children's Hour." *Newsweek,* 16 August 1971, p. 9.

Pollan, Michael. "Reality Shows: The Syndicated Bench." *Channels,* July/August 1987, pp. 52–54.

Sauter, Van Gordon. "In Defense of Tabloid TV." *TV Guide,* 5 August 1989, pp. 2-10.

Sobel, Robert. "Big Glut of Barter Movie Packages; Impact on Syndication Sales Unclear." *Television/Radio Age,* 5 September 1988, pp. 53–54.

_____. "'Oprah' Have-Nots Look for Answers in Early Fringe." *Television/Radio Age,* 12 December 1989, pp. 43–46.

"Syndication at a Glance." *Variety,* 10 January 1990, pp. 79–90.

Tiegel, Eliot. "Vintage Shows: All Legs, No Cash." *Television/Radio Age,* 14 November 1988, pp. 47–49.

Wilson, David S. "Tooned On! America Flips for Cartoons." *TV Guide,* 9 June 1990, pp. 22–29.